1993 YEARBOOK of ASTRONOMY

1993 YEARBOOK of ASTRONOMY

edited by

Patrick Moore

Sidgwick & Jackson Limited
LONDON

First Published 1992 by Sidgwick & Jackson Limited
a division of Pan Macmillan Publishers Limited
Cavaye Place London SW10 9PG
and Basingstoke

Associated companies throughout the world

ISBN 0–283–06096–4 (hardback)
0–283–06097–2 (paperback)

1 2 3 4 5 6 7 8 9

A CIP catalogue record for this book is available from the British Library

Photoset by Rowland Phototypesetting Limited
Bury St Edmunds, Suffolk
Printed in Great Britain by
Butler and Tanner Limited, Frome, Somerset

Contents

Part II: *Article Section*

Part III: *Miscellaneous*

Editor's Foreword

This latest *Yearbook* follows the usual pattern; it seems to have given satisfaction ever since 1962, so that there is no reason to change it! As usual, Gordon Taylor has provided the monthly charts and data, and we have contributions from some of our long-standing contributors as well as newcomers in Jasper Wall, Mark Kidger, Derek McNally and Neil Bone.

David Allen gives us news of an important new instrumental development, while Colin Ronan delves back into history and produces new theories about the invention of the telescope. John Isles has written an account of some particularly interesting variable stars as well as making his usual contribution about the maxima and minima of Mira variables. We also deal with the recent events on Saturn and two subjects nearer home – noctilucent clouds, which bridge astronomy and meteorology, and the ever-present problem of light pollution.

We also include a remembrance to Michael Penston, one of our best-known and best-liked young astronomers, whose death is such a tragic loss to his colleagues all round the world as well as to his many friends.

PATRICK MOORE
Selsey, May 1992

Preface

New readers will find that all the information in this *Yearbook* is given in diagrammatic or descriptive form; the positions of the planets may easily be found on the specially designed star charts, while the monthly notes describe the movements of the planets and give details of other astronomical phenomena visible in both the northern and southern hemispheres. Two sets of the star charts are provided. The **Northern Charts** (pp. 14 to 39) are designed for use in latitude 52 degrees north, but may be used without alteration throughout the British Isles, and (except in the case of eclipses and occultations) in other countries of similar north latitude. The **Southern Charts** (pp. 40 to 65) are drawn for latitude 35 degrees south, and are suitable for use in South Africa, Australia and New Zealand, and other stations in approximately the same south latitude. The reader who needs more detailed information will find *Norton's Star Atlas* (Longman) an invaluable guide, while more precise positions of the planets and their satellites, together with predictions of occultations, meteor showers, and periodic comets may be found in the *Handbook* of the British Astronomical Association. The British monthly periodical, with current news, articles, and monthly notes is *Astronomy Now*. Readers will also find details of forthcoming events given in the American *Sky and Telescope*. This monthly publication also produces a special occultation supplement giving predictions for the United States and Canada.

Important Note
The times given on the star charts and in the Monthly Notes are generally given as local times, using the 24-hour clock, the day beginning at midnight. All the dates, and the times of a few events (e.g. eclipses), are given in Greenwich Mean Time (G.M.T.), which is related to local time by the formula

Local Mean Time = G.M.T. − west longitude

In practice, small differences of longitudes are ignored, and the observer will use local clock time, which will be the appropriate

Standard (or Zone) Time. As the formula indicates, places in west longitude will have a Standard Time slow on G.M.T., while places in east longitude will have a Standard Time fast on G.M.T. As examples we have:

Standard Time in

New Zealand	G.M.T.	+	12 hours
Victoria; N.S.W.	G.M.T.	+	10 hours
Western Australia	G.M.T.	+	8 hours
South Africa	G.M.T.	+	2 hours
British Isles	G.M.T.		
Eastern S.T.	G.M.T.	−	5 hours
Central S.T.	G.M.T.	−	6 hours, etc.

If Summer Time is in use, the clocks will have to have been advanced by one hour, and this hour must be subtracted from the clock time to give Standard Time.

In Great Britain and N. Ireland, Summer Time will be in force in 1993 from March $28^d 01^h$ until October $24^d 01^h$ G.M.T.

Notes on the Star Charts

The stars, together with the Sun, Moon and planets seem to be set on the surface of the celestial sphere, which appears to rotate about the Earth from east to west. Since it is impossible to represent a curved surface accurately on a plane, any kind of star map is bound to contain some form of distortion. But it is well known that the eye can endure some kinds of distortion better than others, and it is particularly true that the eye is most sensitive to deviations from the vertical and horizontal. For this reason the star charts given in this volume have been designed to give a true representation of vertical and horizontal lines, whatever may be the resulting distortion in the shape of a constellation figure. It will be found that the amount of distortion is, in general, quite small, and is only obvious in the case of large constellations such as Leo and Pegasus, when these appear at the top of the charts, and so are drawn out sideways.

The charts show all stars down to the fourth magnitude, together with a number of fainter stars which are necessary to define the shape of a constellation. There is no standard system for representing the outlines of the constellations, and triangles and other simple figures have been used to give outlines which are easy to follow with the naked eye. The names of the constellations are given, together with the proper names of the brighter stars. The apparent magnitudes of the stars are indicated roughly by using four different sizes of dots, the larger dots representing the brighter stars.

The two sets of star charts are similar in design. At each opening there is a group of four charts which give a complete coverage of the sky up to an altitude of 62½ degrees; there are twelve such groups to cover the entire year. In the **Northern Charts** (for 52 degrees north) the upper two charts show the southern sky, south being at the centre and east on the left. The coverage is from 10 degrees north of east (top left) to 10 degrees north of west (top right). The two lower charts show the northern sky from 10 degrees south of west (lower left) to 10 degrees south of east (lower right). There is thus an overlap east and west.

Conversely, in the **Southern Charts** (for 35 degrees south) the upper two charts show the northern sky, with north at the centre and east on the right. The two lower charts show the southern sky, with south at the centre and east on the left. The coverage and overlap is the same on both sets of charts.

Because the sidereal day is shorter than the solar day, the stars appear to rise and set about four minutes earlier each day, and this amounts to two hours in a month. Hence the twelve groups of charts in each set are sufficient to give the appearance of the sky throughout the day at intervals of two hours, or at the same time of night at monthly intervals throughout the year. The actual range of dates and times when the stars on the charts are visible is indicated at the top of each page. Each group is numbered in bold type, and the number to be used for any given month and time is summarized in the following table:

Local Time	18h	20h	22h	0h	2h	4h	6h
January	11	12	1	2	3	4	5
February	12	1	2	3	4	5	6
March	1	2	3	4	5	6	7
April	2	3	4	5	6	7	8
May	3	4	5	6	7	8	9
June	4	5	6	7	8	9	10
July	5	6	7	8	9	10	11
August	6	7	8	9	10	11	12
September	7	8	9	10	11	12	1
October	8	9	10	11	12	1	2
November	9	10	11	12	1	2	3
December	10	11	12	1	2	3	4

The charts are drawn to scale, the horizontal measurements, marked at every 10 degrees, giving the azimuths (or true bearings) measured from the north round through east (90 degrees), south (180 degrees), and west (270 degrees). The vertical measurements, similarly marked, give the altitudes of the stars up to 62½ degrees. Estimates of altitude and azimuth made from these charts will necessarily be mere approximations, since no observer will be exactly at the adopted latitude, or at the stated time, but they will serve for the identification of stars and planets.

The ecliptic is drawn as a broken line on which longitude is marked at every 10 degrees; the positions of the planets are then easily found by reference to the table on page 71. It will be noticed that on the Southern Charts the **ecliptic** may reach an altitude in excess of 62½ degrees on star charts 5 to 9. The continuations of the broken line will be found on the charts of overhead stars.

There is a curious illusion that stars at an altitude of 60 degrees or more are actually overhead, and the beginner may often feel that he is leaning over backwards in trying to see them. These overhead stars are given separately on the pages immediately following the main star charts. The entire year is covered at one opening, each of the four maps showing the overhead stars at times which correspond to those of three of the main star charts. The position of the zenith is indicated by a cross, and this cross marks the centre of a circle which is 35 degrees from the zenith; there is thus a small overlap with the main charts.

The broken line leading from the north (on the Northern Charts) or from the south (on the Southern Charts) is numbered to indicate the corresponding main chart. Thus on page 38 the N-S line numbered 6 is to be regarded as an extension of the centre (south) line of chart 6 on pages 24 and 25, and at the top of these pages are printed the dates and times which are appropriate. Similarly, on page 65, the S-N line numbered 10 connects with the north line of the upper charts on pages 58 and 59.

The overhead stars are plotted as maps on a conical projection, and the scale is rather smaller than that of the main charts.

1L

October 6 at	5^h	October 21 at	4^h
November 6 at	3^h	November 21 at	2^h
December 6 at	1^h	December 21 at	midnight
January 6 at	23^h	January 21 at	22^h
February 6 at	21^h	February 21 at	20^h

October 6 at 5^h	October 21 at 4^h	**1R**
November 6 at 3^h	November 21 at 2^h	
December 6 at 1^h	December 21 at midnight	
January 6 at 23^h	January 21 at 22^h	
February 6 at 21^h	February 21 at 20^h	

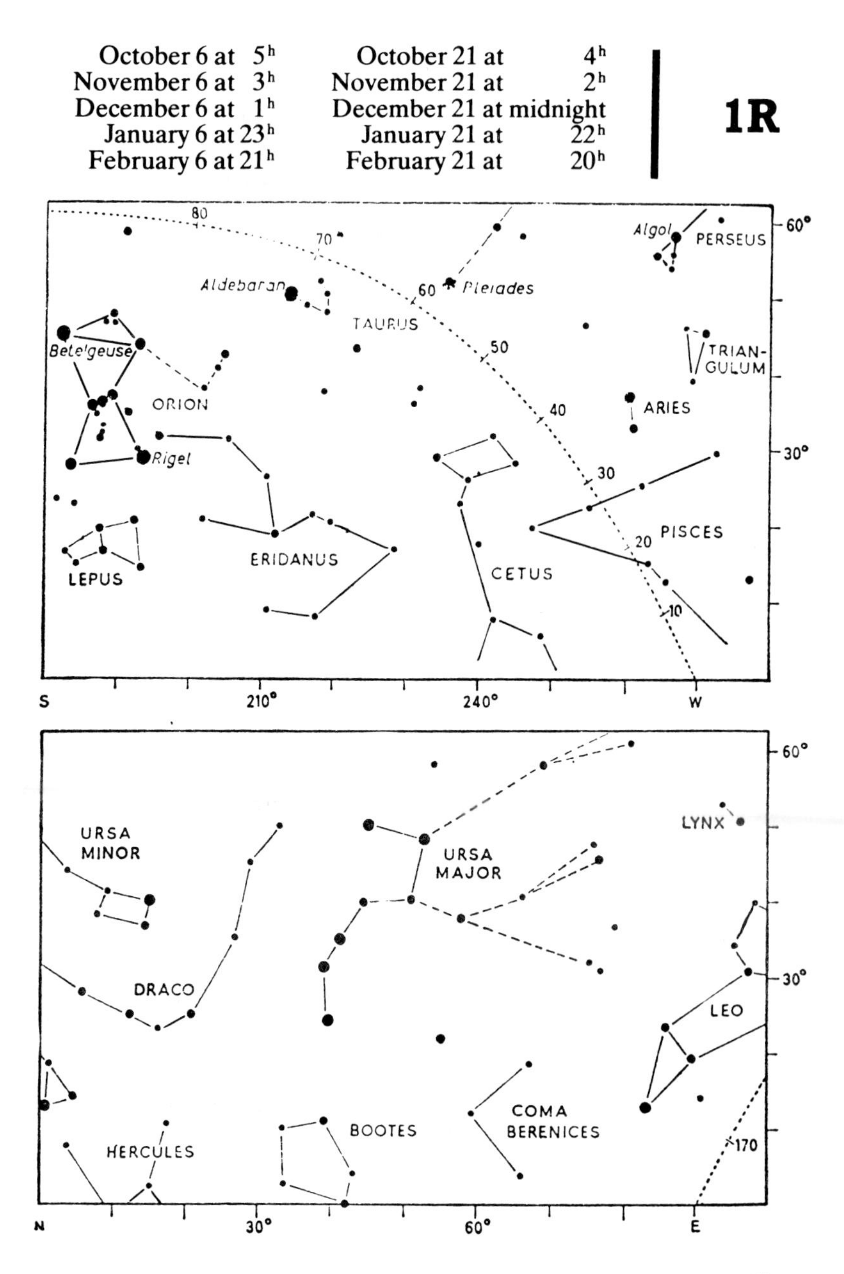

2L

November 6 at 5h	November 21 at 4h
December 6 at 3h	December 21 at 2h
January 6 at 1h	January 21 at midnight
February 6 at 23h	February 21 at 22h
March 6 at 21h	March 21 at 20h

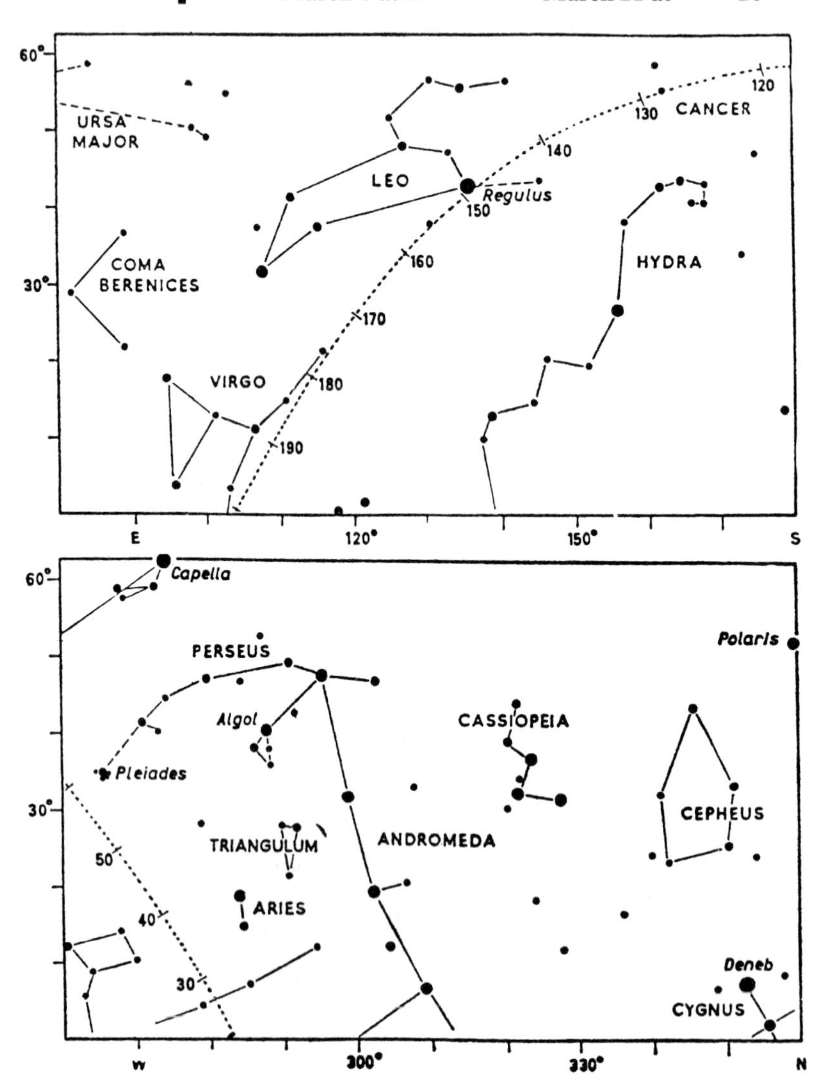

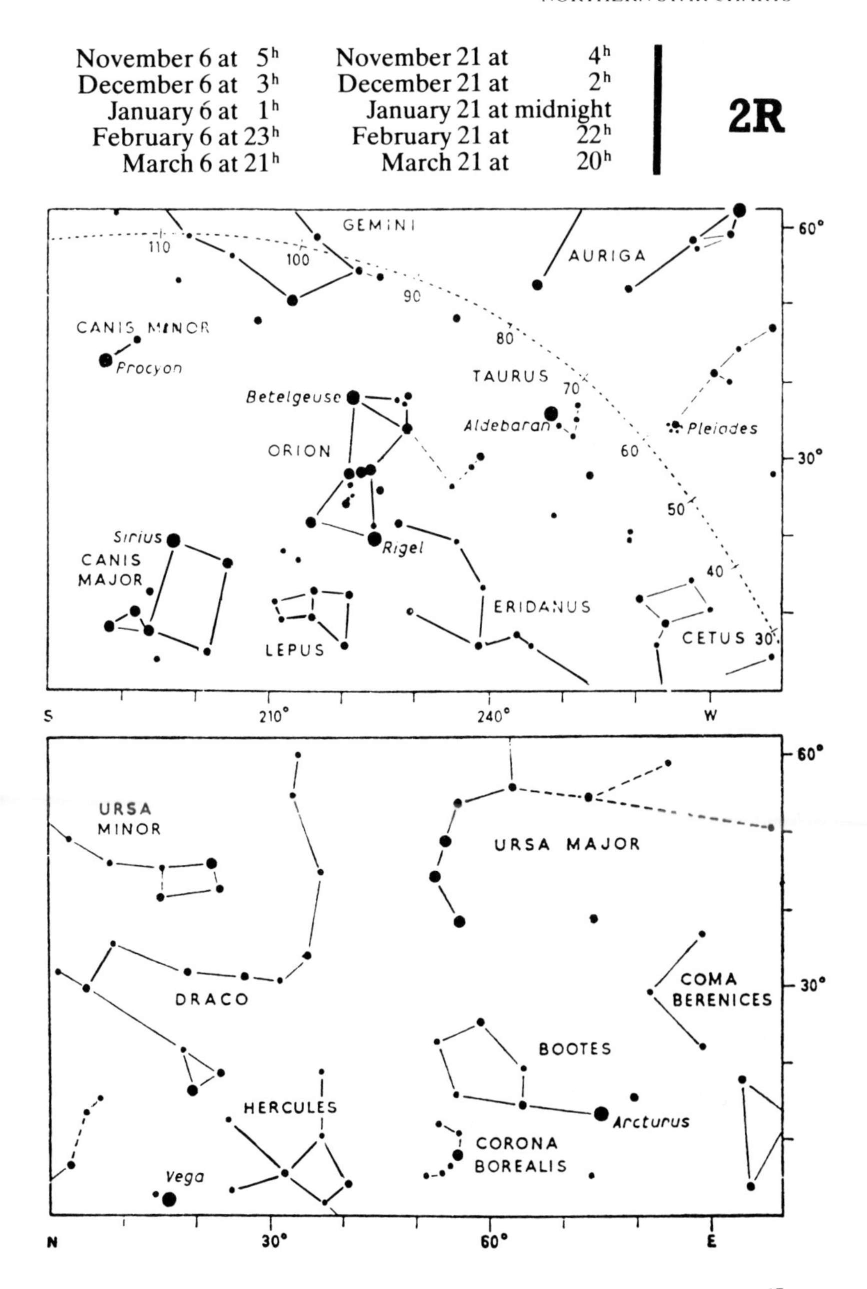
November 6 at 5h
December 6 at 3h
January 6 at 1h
February 6 at 23h
March 6 at 21h
November 21 at 4h
December 21 at 2h
January 21 at midnight
February 21 at 22h
March 21 at 20h
2R
GEMINI
AURIGA
110
100
90
80
70
60
50
40
30
CANIS MINOR
Procyon
TAURUS
Betelgeuse
Aldebaran
Pleiades
ORION
Sirius
CANIS MAJOR
Rigel
ERIDANUS
LEPUS
CETUS
60°
30°
S
210°
240°
W
URSA MINOR
URSA MAJOR
DRACO
COMA BERENICES
BOOTES
HERCULES
Arcturus
CORONA BOREALIS
Vega
60°
30°
N
30°
60°
E

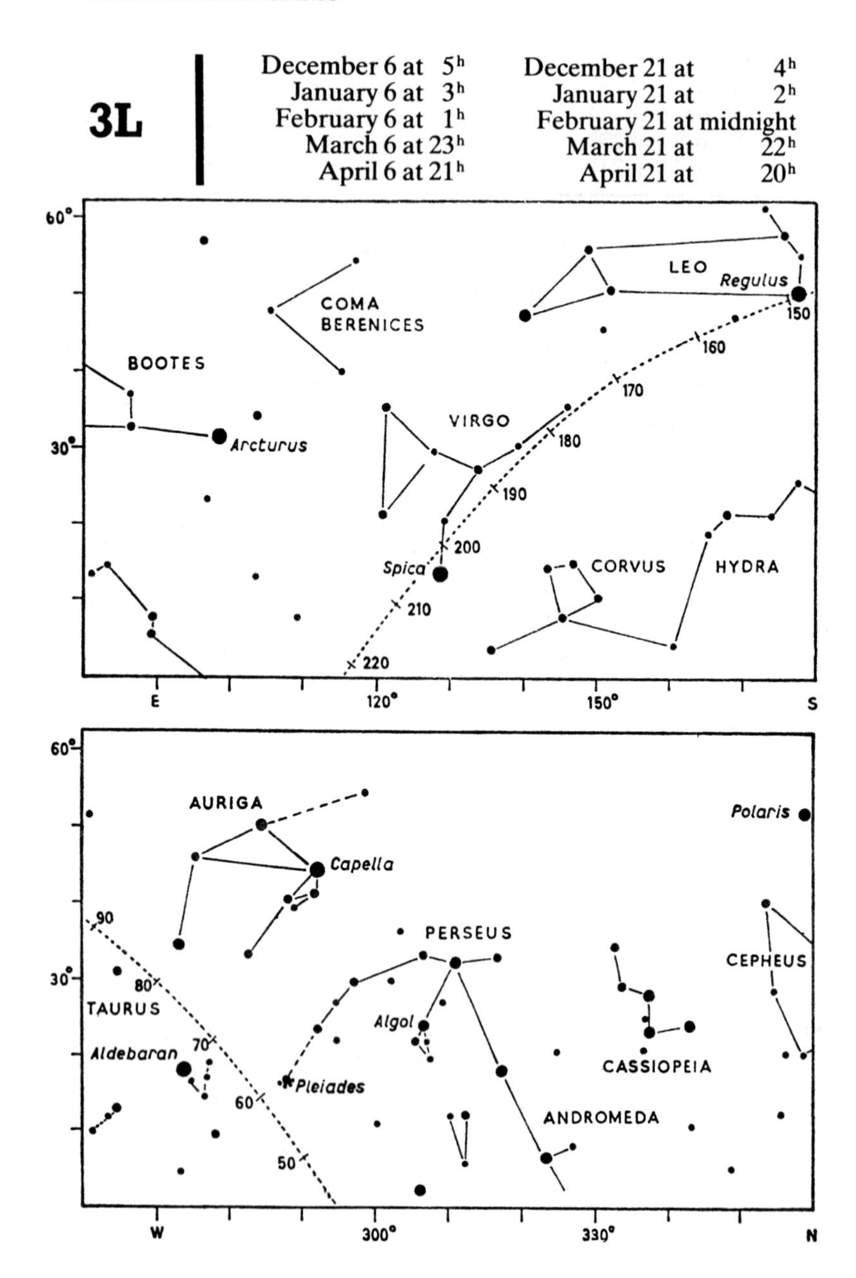
3L
December 6 at 5h
January 6 at 3h
February 6 at 1h
March 6 at 23h
April 6 at 21h
December 21 at 4h
January 21 at 2h
February 21 at midnight
March 21 at 22h
April 21 at 20h
60°
30°
LEO
Regulus
COMA
BERENICES
BOOTES
Arcturus
VIRGO
Spica
CORVUS
HYDRA
150
160
170
180
190
200
210
220
E
120°
150°
S
AURIGA
Capella
Polaris
PERSEUS
CEPHEUS
TAURUS
Aldebaran
Algol
Pleiades
CASSIOPEIA
ANDROMEDA
90
80
70
60
50
W
300°
330°
N

December 6 at 5h	December 21 at 4h
January 6 at 3h	January 21 at 2h
February 6 at 1h	February 21 at midnight
March 6 at 23h	March 21 at 22h
April 6 at 21h	April 21 at 20h

3R

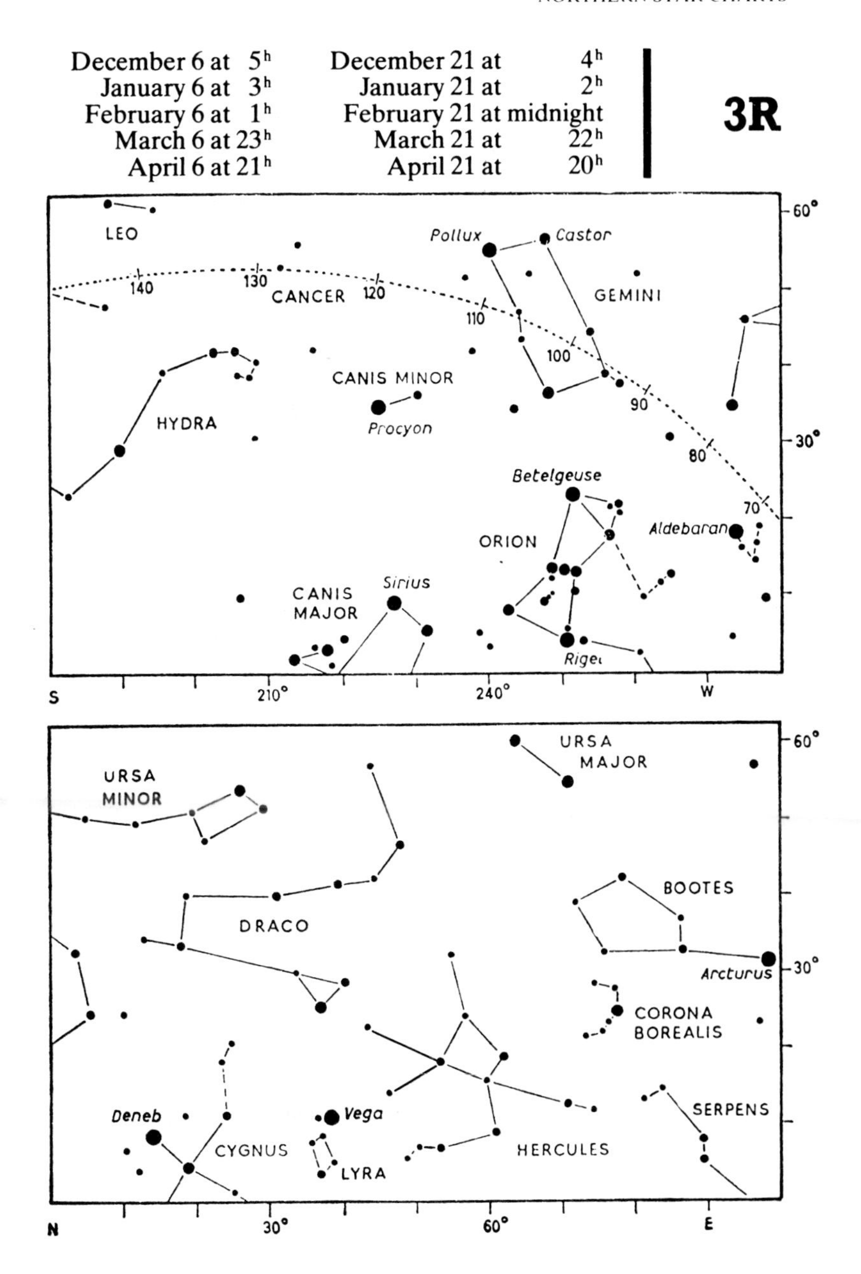

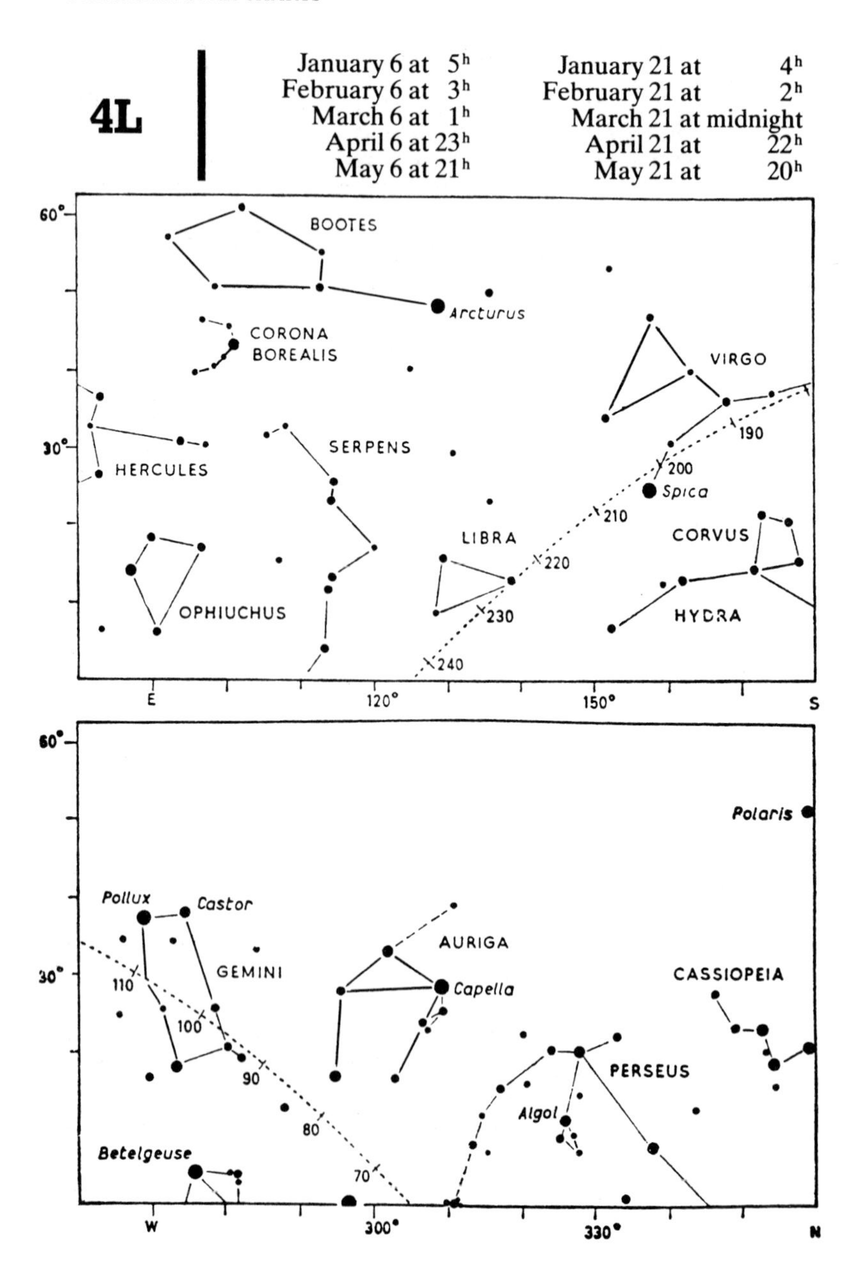
4L
January 6 at 5h
February 6 at 3h
March 6 at 1h
April 6 at 23h
May 6 at 21h
January 21 at 4h
February 21 at 2h
March 21 at midnight
April 21 at 22h
May 21 at 20h
60°
30°
BOOTES
Arcturus
CORONA
BOREALIS
VIRGO
HERCULES
SERPENS
190
200
Spica
210
LIBRA
CORVUS
220
OPHIUCHUS
230
HYDRA
240
E
120°
150°
S
60°
Polaris
Pollux
Castor
AURIGA
30°
GEMINI
CASSIOPEIA
110
Capella
100
90
PERSEUS
Algol
80
Betelgeuse
70
W
300°
330°
N

January 6 at 5ʰ	January 21 at 4ʰ	**4R**
February 6 at 3ʰ	February 21 at 2ʰ	
March 6 at 1ʰ	March 21 at midnight	
April 6 at 23ʰ	April 21 at 22ʰ	
May 6 at 21ʰ	May 21 at 20ʰ	

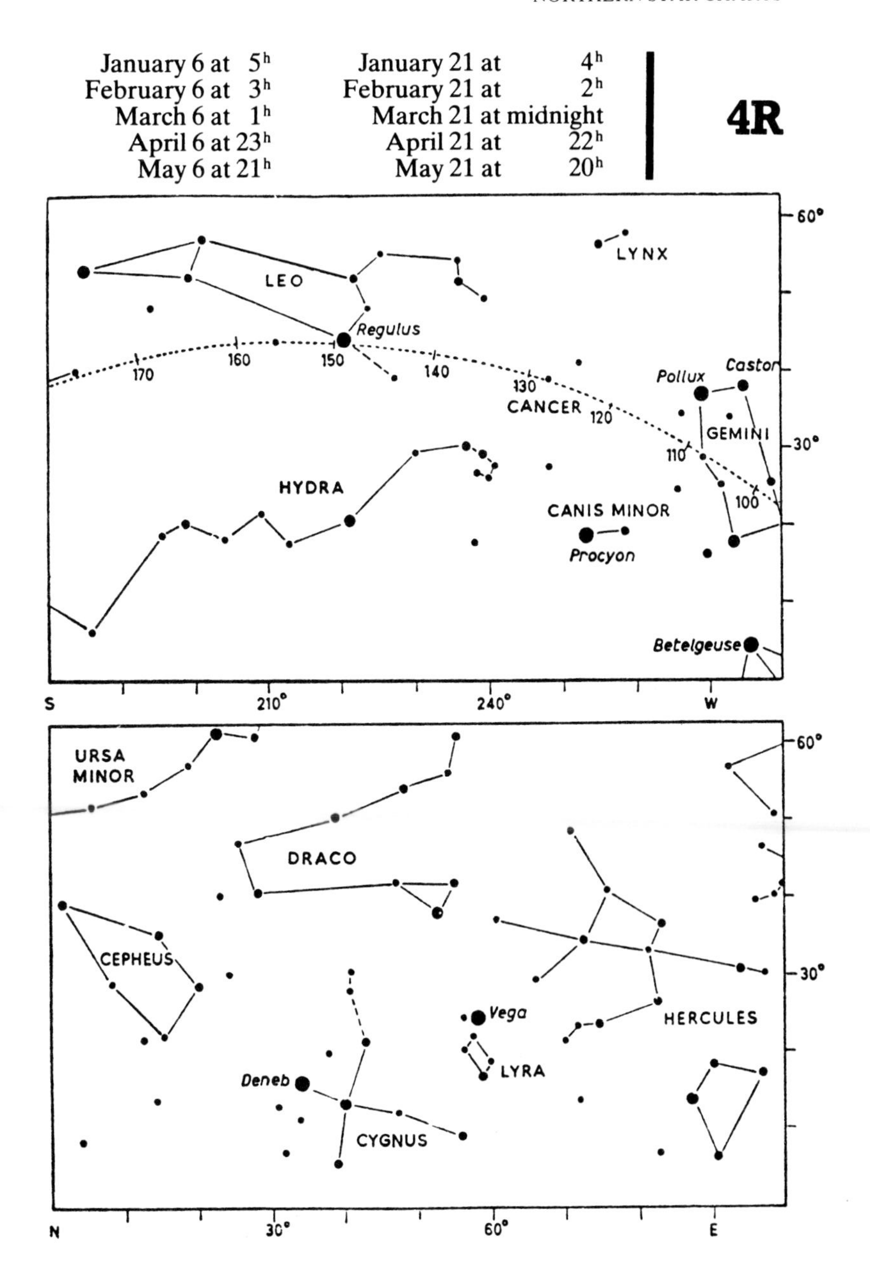

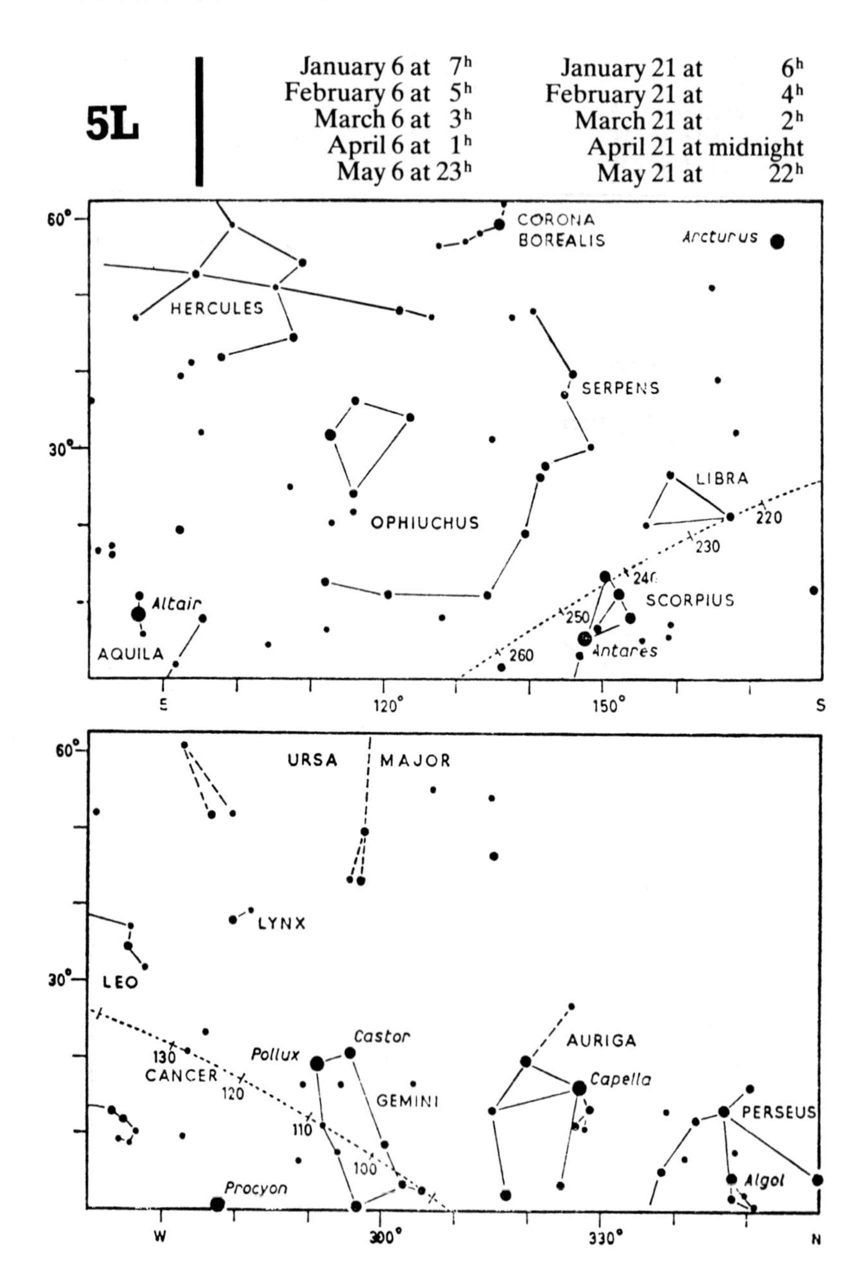
5L
January 6 at 7h
February 6 at 5h
March 6 at 3h
April 6 at 1h
May 6 at 23h
January 21 at 6h
February 21 at 4h
March 21 at 2h
April 21 at midnight
May 21 at 22h
60°
30°
CORONA BOREALIS
Arcturus
HERCULES
SERPENS
OPHIUCHUS
LIBRA
220
230
240
250
260
SCORPIUS
Antares
Altair
AQUILA
E
120°
150°
S
URSA MAJOR
LYNX
LEO
130
CANCER
120
110
100
Pollux
Castor
GEMINI
AURIGA
Capella
PERSEUS
Algol
Procyon
W
300°
330°
N

January 6 at 7ʰ	January 21 at 6ʰ	**5R**
February 6 at 5ʰ	February 21 at 4ʰ	
March 6 at 3ʰ	March 21 at 2ʰ	
April 6 at 1ʰ	April 21 at midnight	
May 6 at 23ʰ	May 21 at 22ʰ	

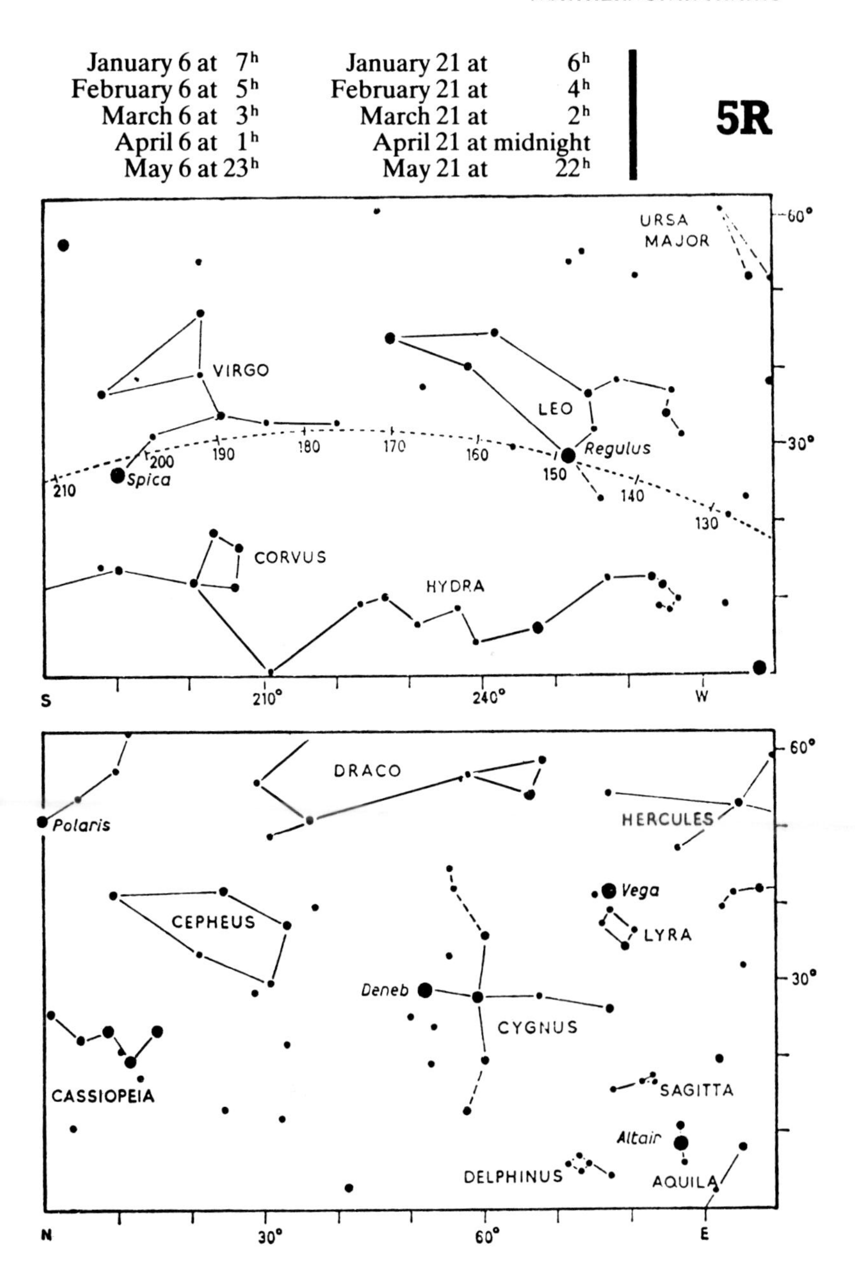

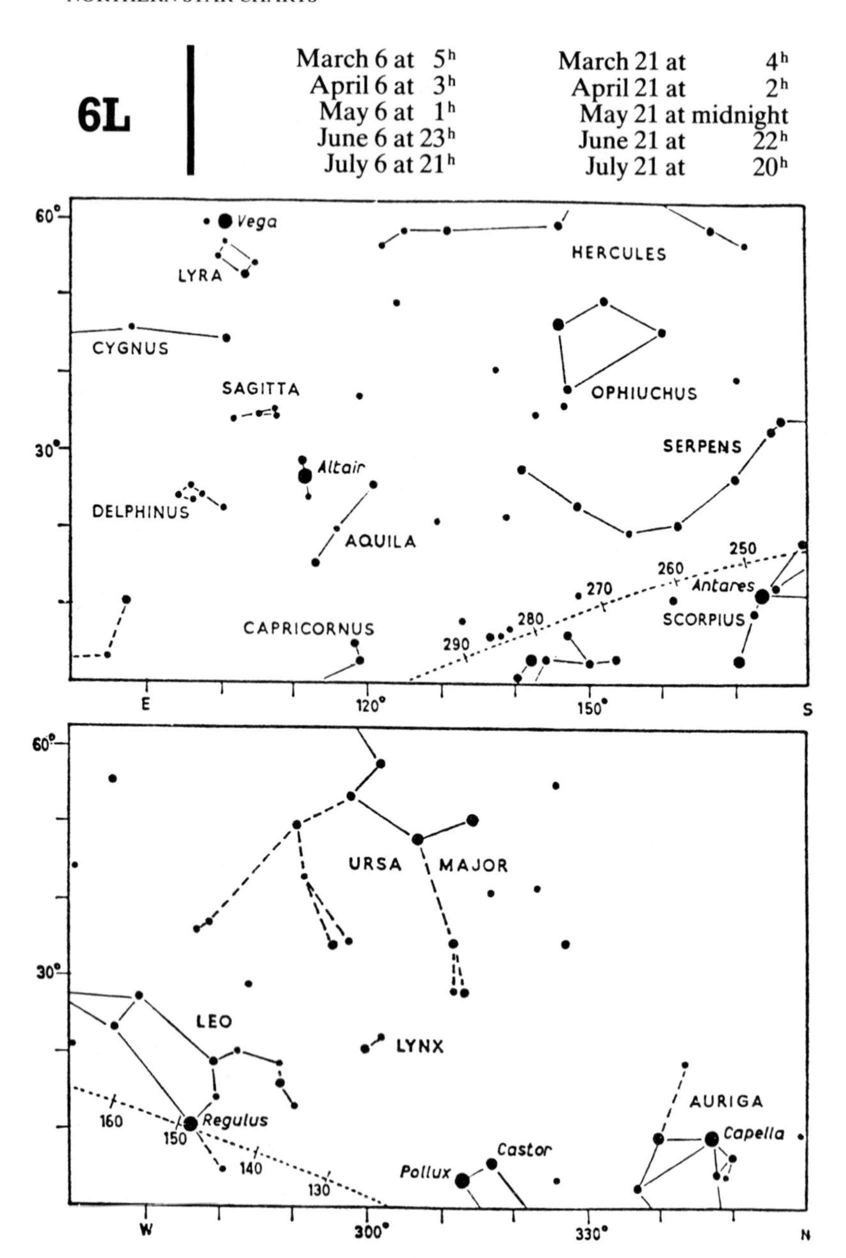
6L
March 6 at 5h
April 6 at 3h
May 6 at 1h
June 6 at 23h
July 6 at 21h
March 21 at 4h
April 21 at 2h
May 21 at midnight
June 21 at 22h
July 21 at 20h
60°
30°
Vega
LYRA
HERCULES
CYGNUS
SAGITTA
OPHIUCHUS
SERPENS
Altair
DELPHINUS
AQUILA
250
260
270
Antares
280
SCORPIUS
CAPRICORNUS
290
E
120°
150°
S
60°
URSA
MAJOR
30°
LEO
LYNX
AURIGA
160
Regulus
150
Capella
Castor
140
Pollux
130
W
300°
330°
N

March 6 at 5^h	March 21 at 4^h	**6R**
April 6 at 3^h	April 21 at 2^h	
May 6 at 1^h	May 21 at midnight	
June 6 at 23^h	June 21 at 22^h	
July 6 at 21^h	July 21 at 20^h	

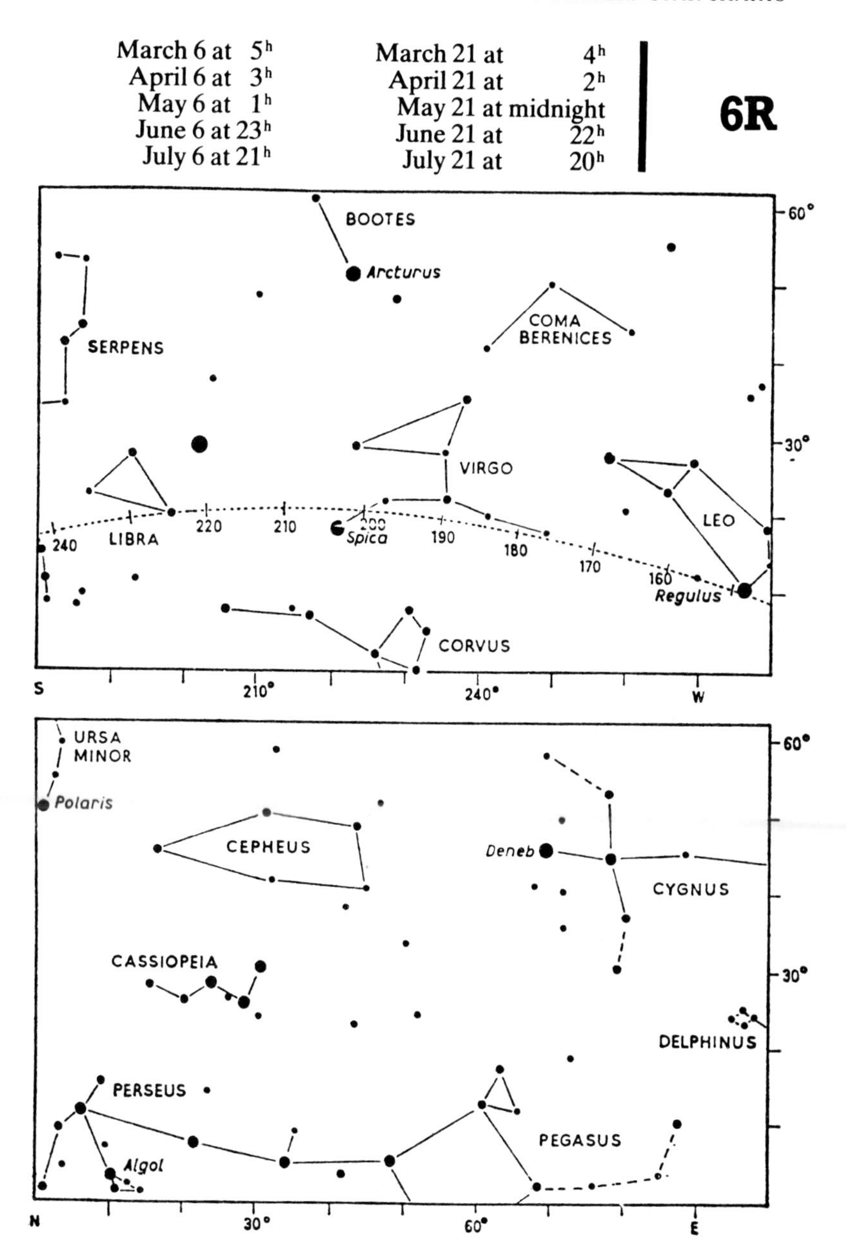

7L

May 6 at 3ʰ	May 21 at 2ʰ
June 6 at 1ʰ	June 21 at midnight
July 6 at 23ʰ	July 21 at 22ʰ
August 6 at 21ʰ	August 21 at 20ʰ
September 6 at 19ʰ	September 21 at 18ʰ

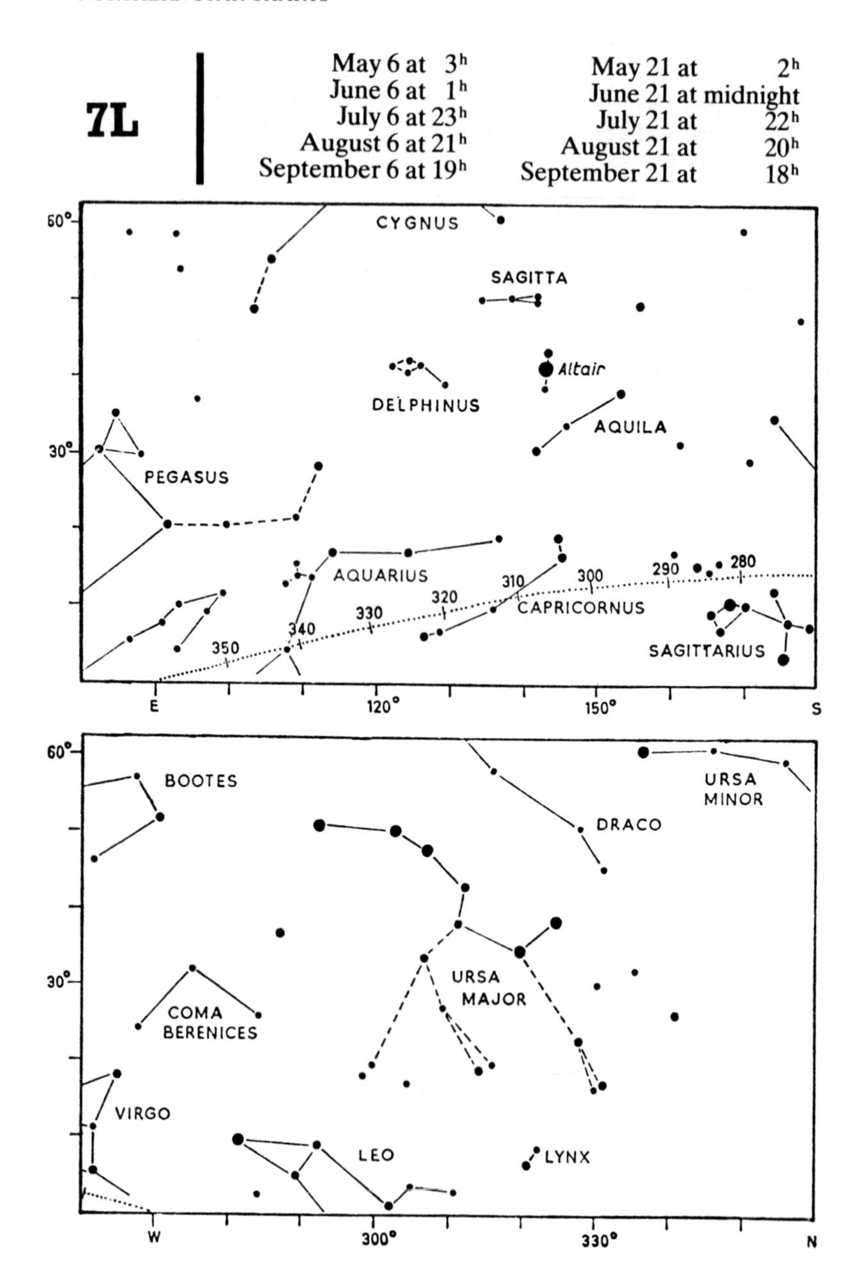

May 6 at 3h	May 21 at 2h	**7R**
June 6 at 1h	June 21 at midnight	
July 6 at 23h	July 21 at 22h	
August 6 at 21h	August 21 at 20h	
September 6 at 19h	September 21 at 18h	

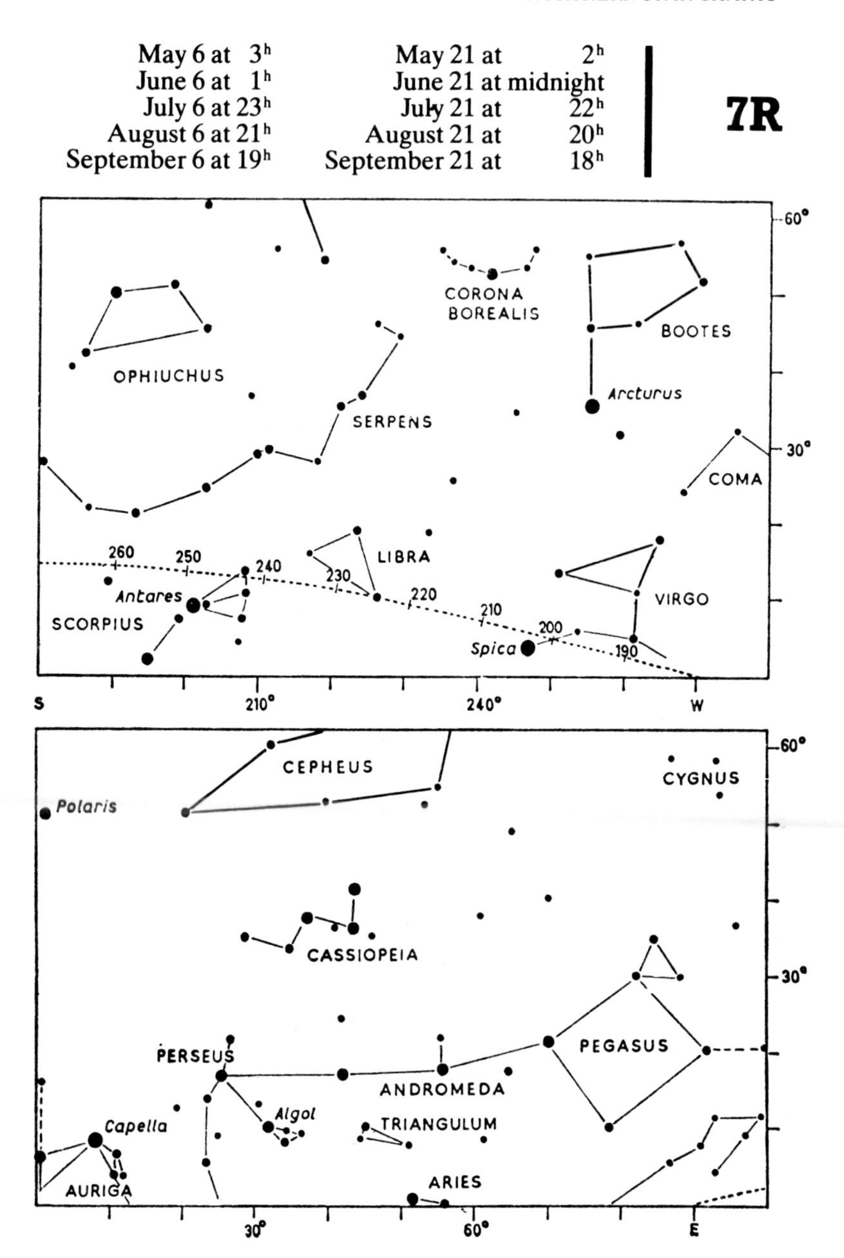

8L

July 6 at 1^h	July 21 at midnight
August 6 at 23^h	August 21 at 22^h
September 6 at 21^h	September 21 at 20^h
October 6 at 19^h	October 21 at 18^h
November 6 at 17^h	November 21 at 16^h

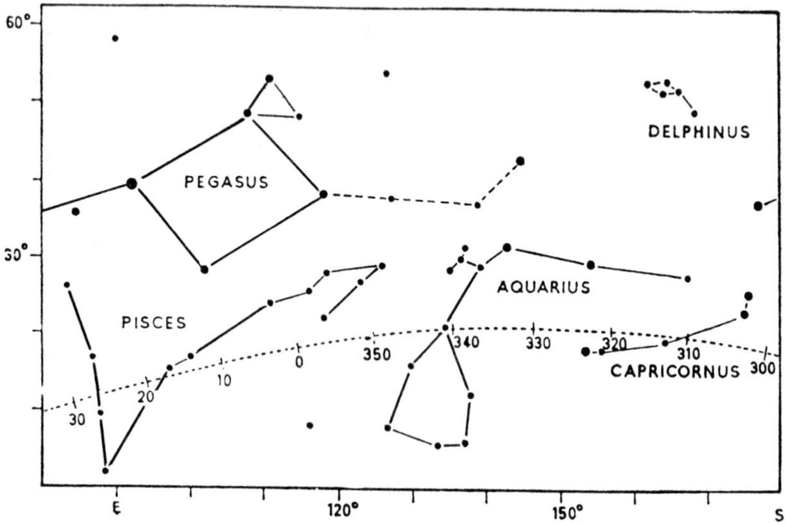

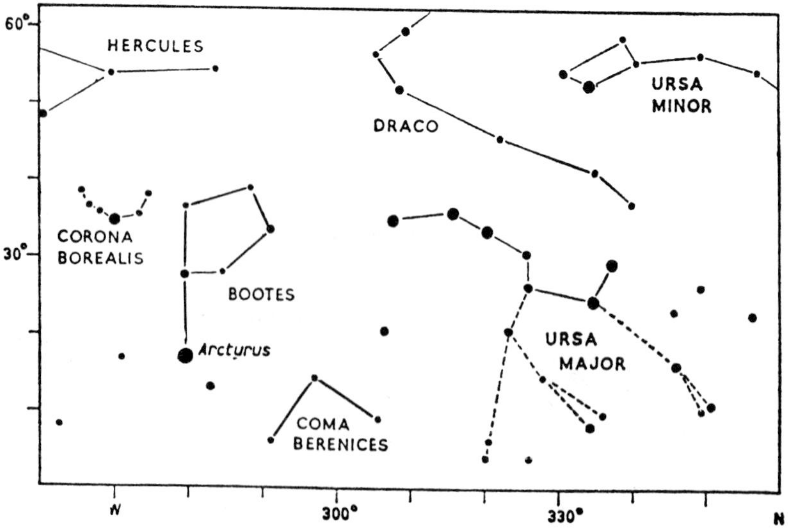

July 6 at 1h	July 21 at midnight	**8R**
August 6 at 23h	August 21 at 22h	
September 6 at 21h	September 21 at 20h	
October 6 at 19h	October 21 at 18h	
November 6 at 17h	November 21 at 16h	

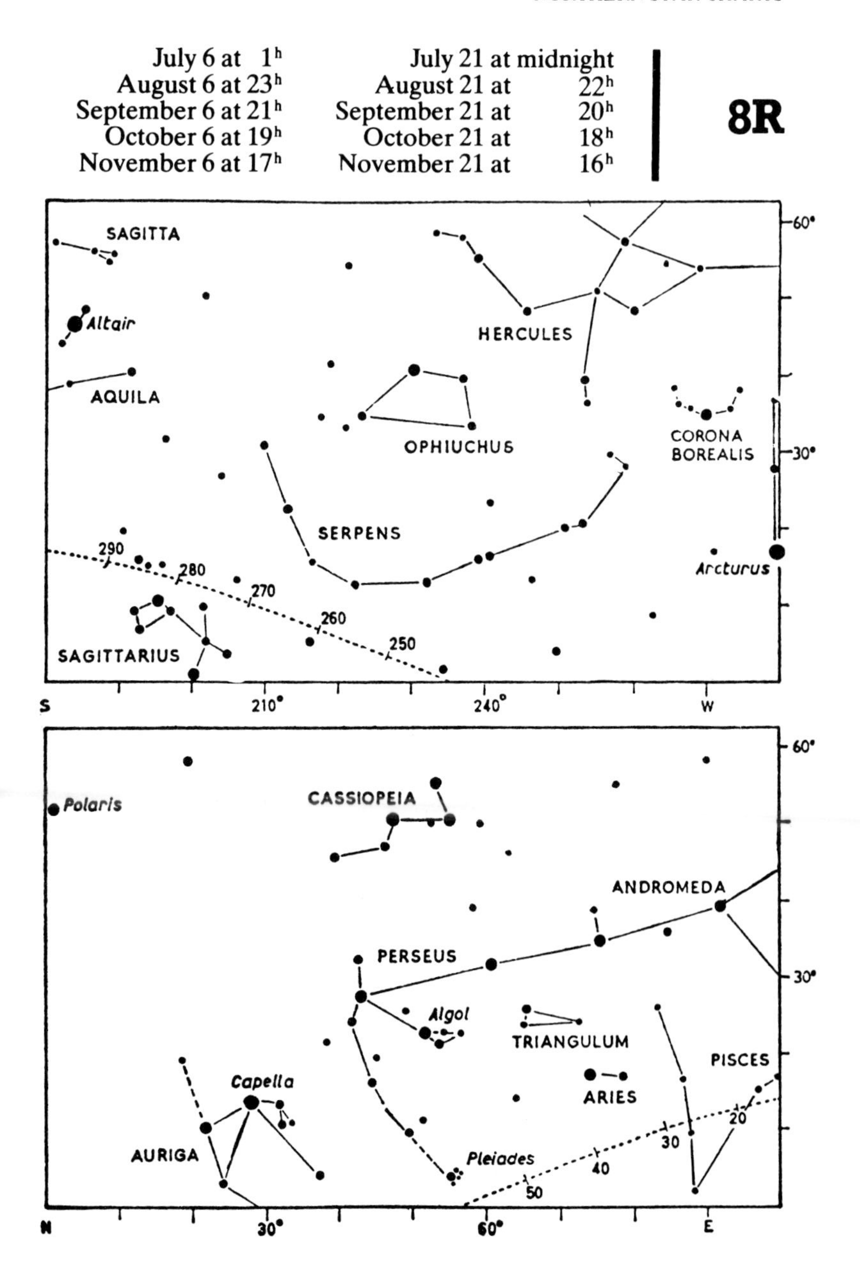

9L

August 6 at 1^h	August 21 at midnight
September 6 at 23^h	September 21 at 22^h
October 6 at 21^h	October 21 at 20^h
November 6 at 19^h	November 21 at 18^h
December 6 at 17^h	December 21 at 16^h

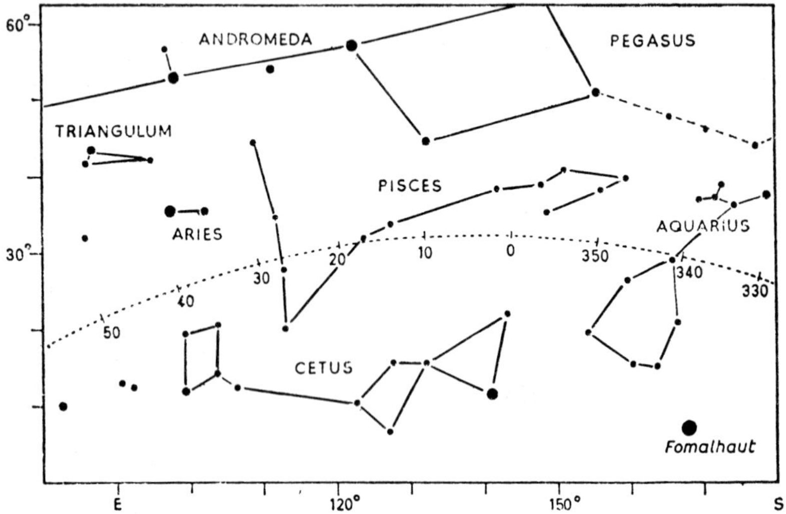

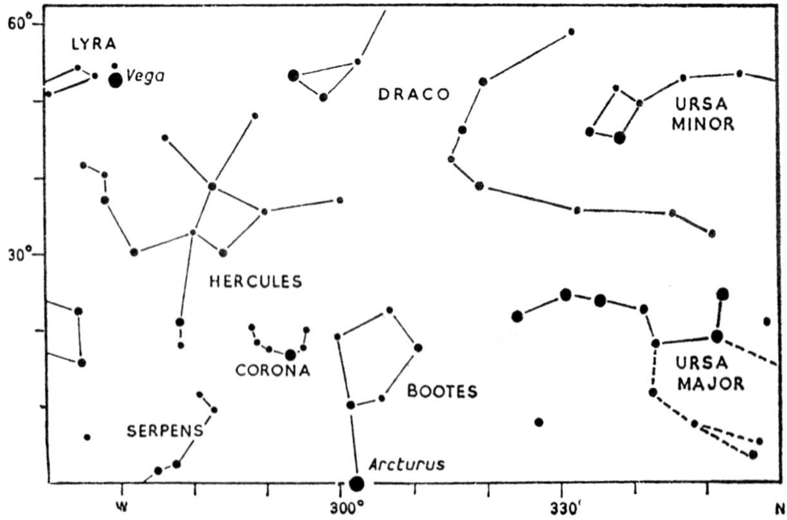

60°
ANDROMEDA
PEGASUS
TRIANGULUM
PISCES
ARIES
AQUARIUS
30°
20
10
0
350
340
330
30
40
50
CETUS
Fomalhaut
E
120°
150°
S

60°
LYRA
Vega
DRACO
URSA MINOR
30°
HERCULES
CORONA
BOOTES
URSA MAJOR
SERPENS
Arcturus
W
300°
330°
N

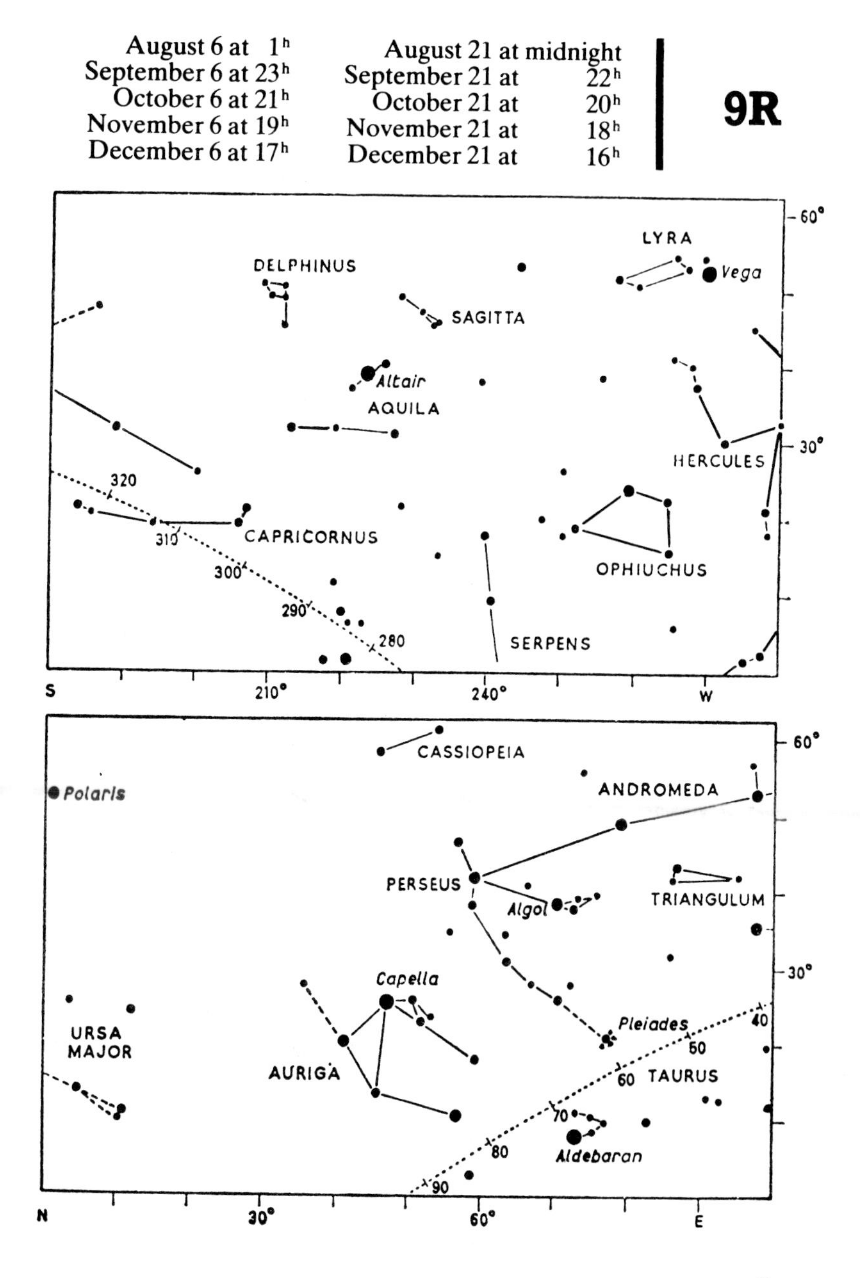
August 6 at 1h
September 6 at 23h
October 6 at 21h
November 6 at 19h
December 6 at 17h
August 21 at midnight
September 21 at 22h
October 21 at 20h
November 21 at 18h
December 21 at 16h
9R
LYRA
Vega
DELPHINUS
SAGITTA
Altair
AQUILA
HERCULES
CAPRICORNUS
OPHIUCHUS
SERPENS
320
310
300
290
280
60°
30°
S
210°
240°
W
CASSIOPEIA
Polaris
ANDROMEDA
PERSEUS
Algol
TRIANGULUM
Capella
Pleiades
URSA MAJOR
AURIGA
TAURUS
Aldebaran
40
50
60
70
80
90
60°
30°
N
30°
60°
E

10L

August 6 at 3h	August 21 at 2h
September 6 at 1h	September 21 at midnight
October 6 at 23h	October 21 at 22h
November 6 at 21h	November 21 at 20h
December 6 at 19h	December 21 at 18h

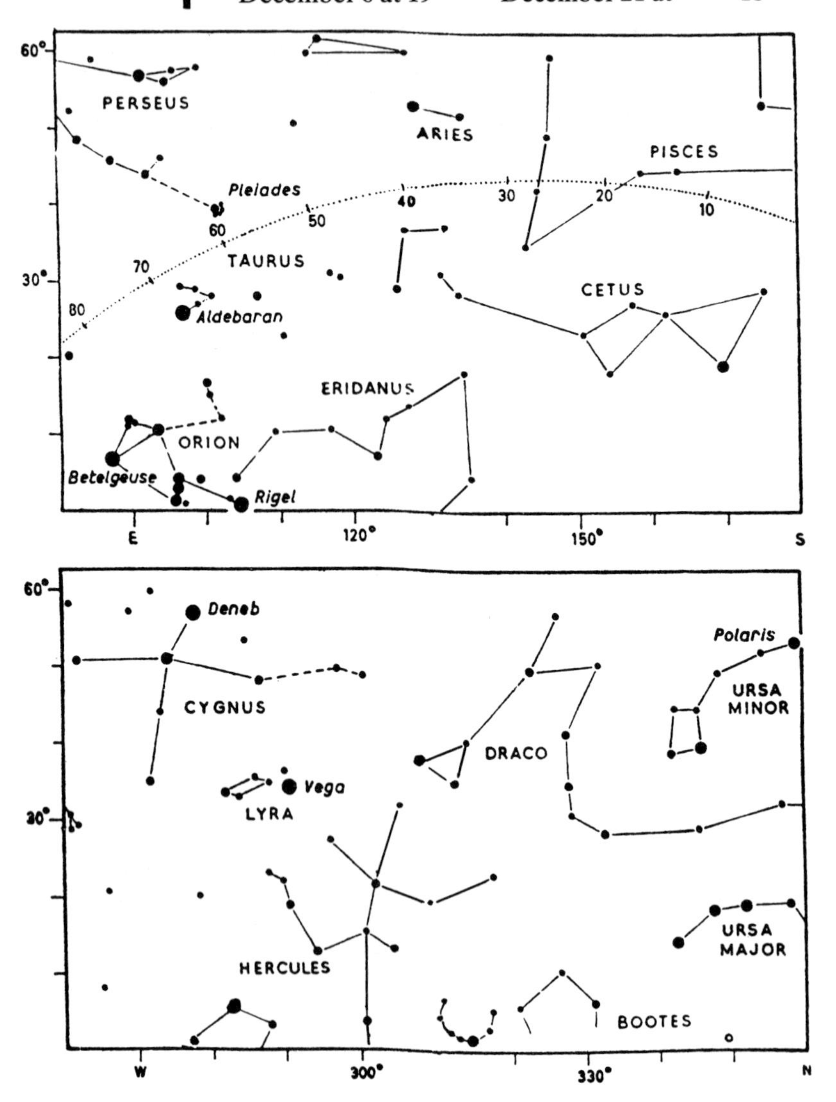

August 6 at 3h	August 21 at 2h	**10R**
September 6 at 1h	September 21 at midnight	
October 6 at 23h	October 21 at 22h	
November 6 at 21h	November 21 at 20h	
December 6 at 19h	December 21 at 18h	

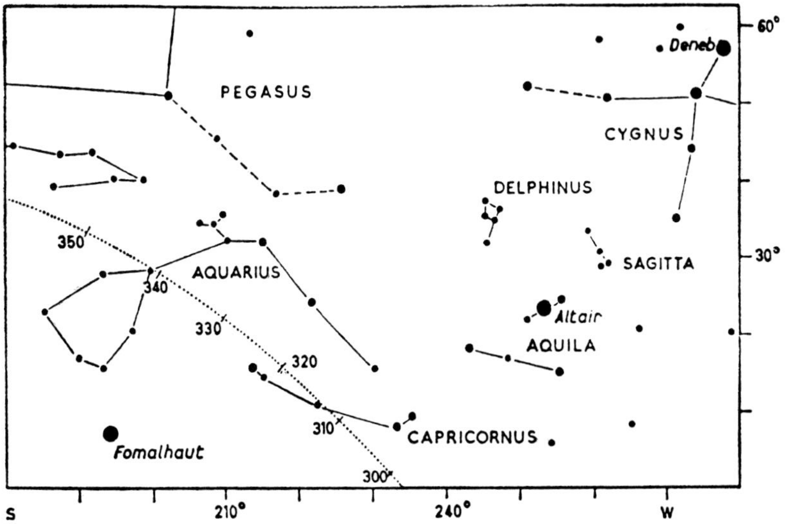

PEGASUS
CYGNUS
Deneb
DELPHINUS
SAGITTA
AQUARIUS
Altair
AQUILA
CAPRICORNUS
Fomalhaut
350
340
330
320
310
300
60°
30°
S
210°
240°
W

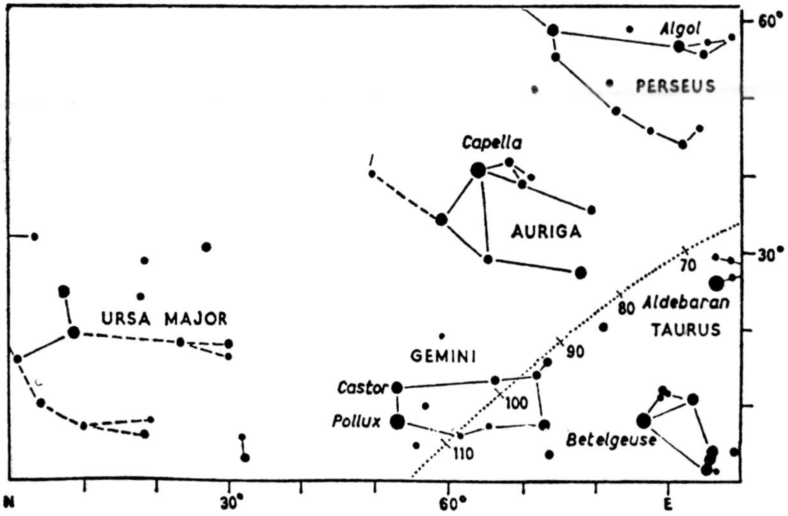

Algol
PERSEUS
Capella
AURIGA
URSA MAJOR
Aldebaran
TAURUS
GEMINI
Castor
Pollux
Betelgeuse
70
80
90
100
110
60°
30°
N
30°
60°
E

11L

September 6 at 3h	September 21 at 2h
October 6 at 1h	October 21 at midnight
November 6 at 23h	November 21 at 22h
December 6 at 21h	December 21 at 20h
January 6 at 19h	January 21 at 18h

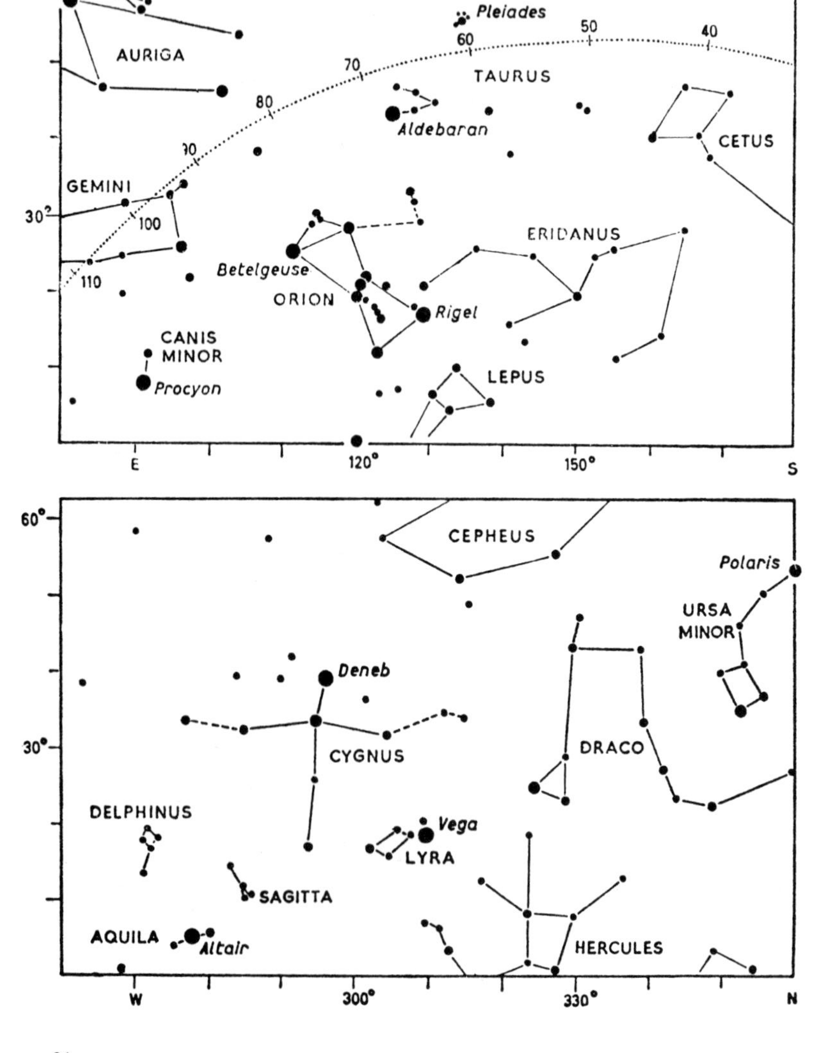

September 6 at 3ʰ	September 21 at 2ʰ	
October 6 at 1ʰ	October 21 at midnight	
November 6 at 23ʰ	November 21 at 22ʰ	**11R**
December 6 at 21ʰ	December 21 at 20ʰ	
January 6 at 19ʰ	January 21 at 18ʰ	

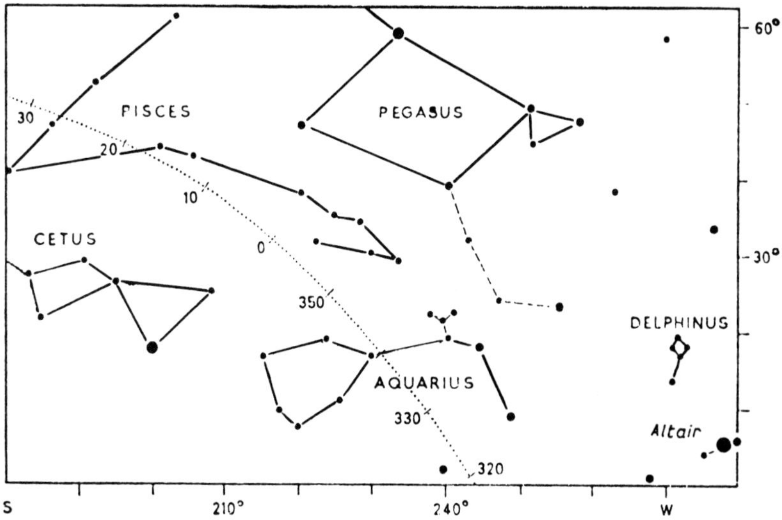

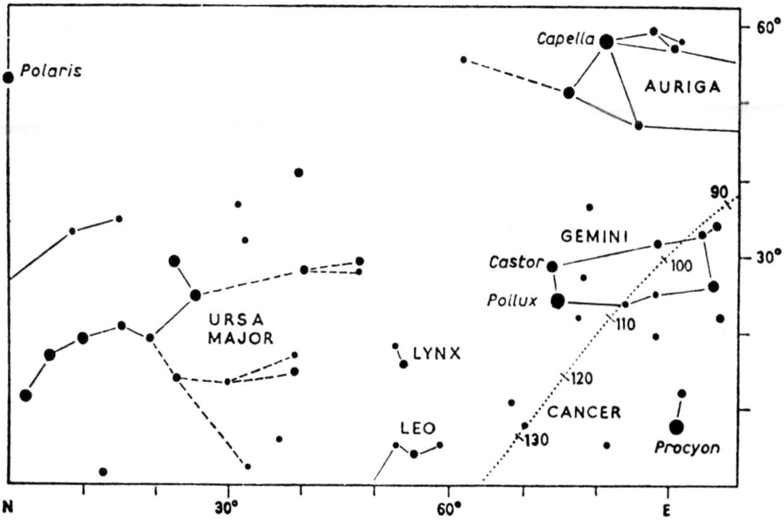

12L

October 6 at 3h	October 21 at 2h
November 6 at 1h	November 21 at midnight
December 6 at 23h	December 21 at 22h
January 6 at 21h	January 21 at 20h
February 6 at 19h	February 21 at 18h

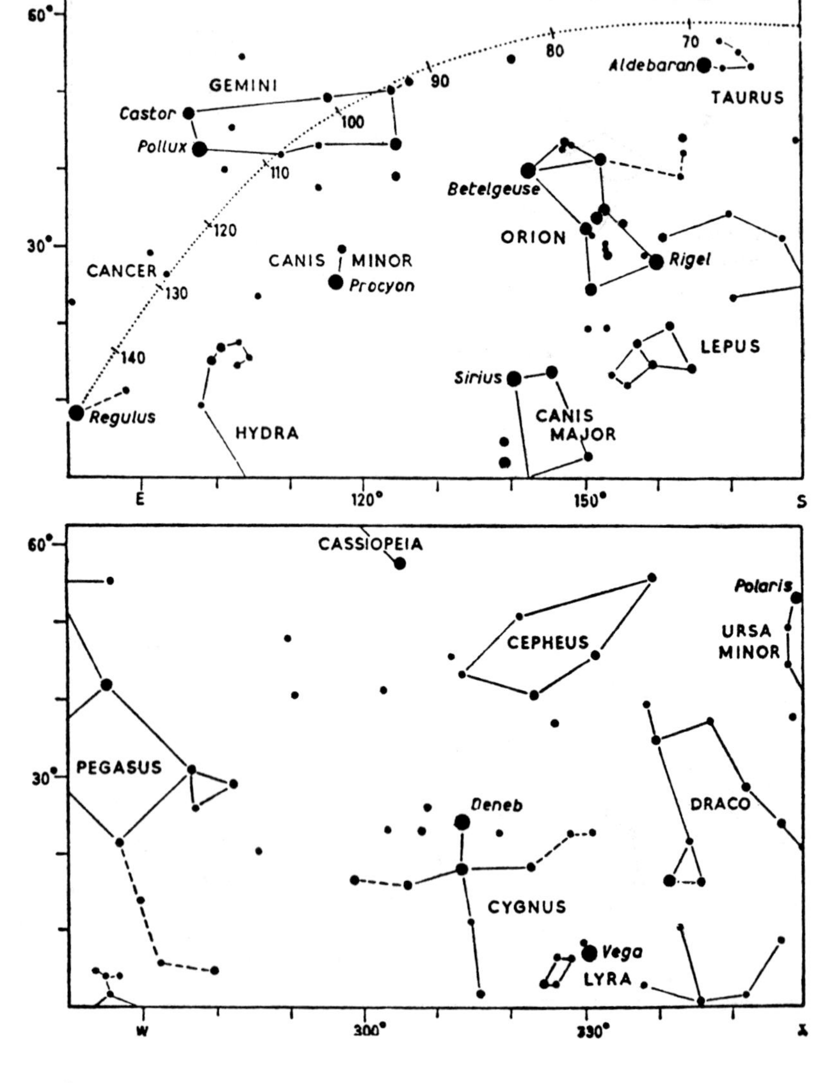

October 6 at 3h	October 21 at 2h	**12R**
November 6 at 1h	November 21 at midnight	
December 6 at 23h	December 21 at 22h	
January 6 at 21h	January 21 at 20h	
February 6 at 19h	February 21 at 18h	

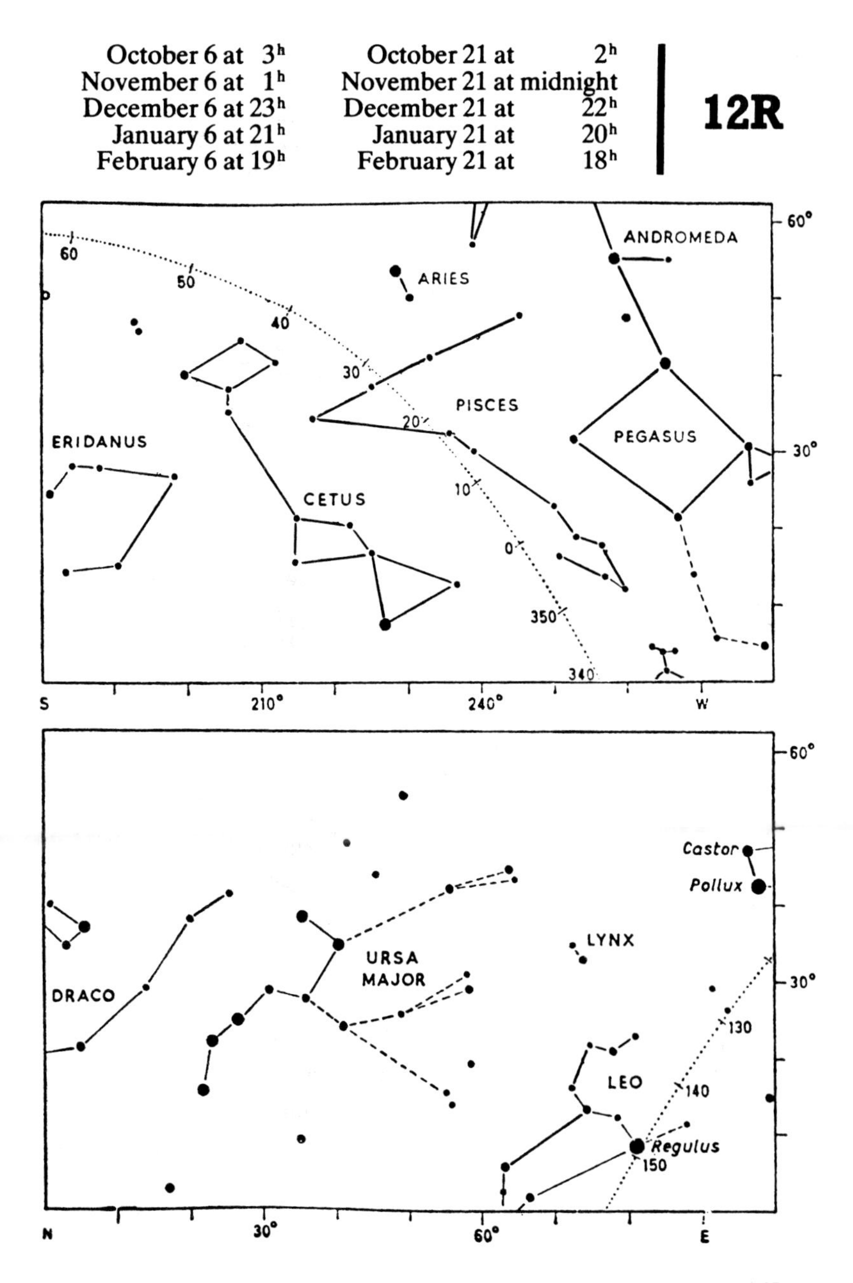

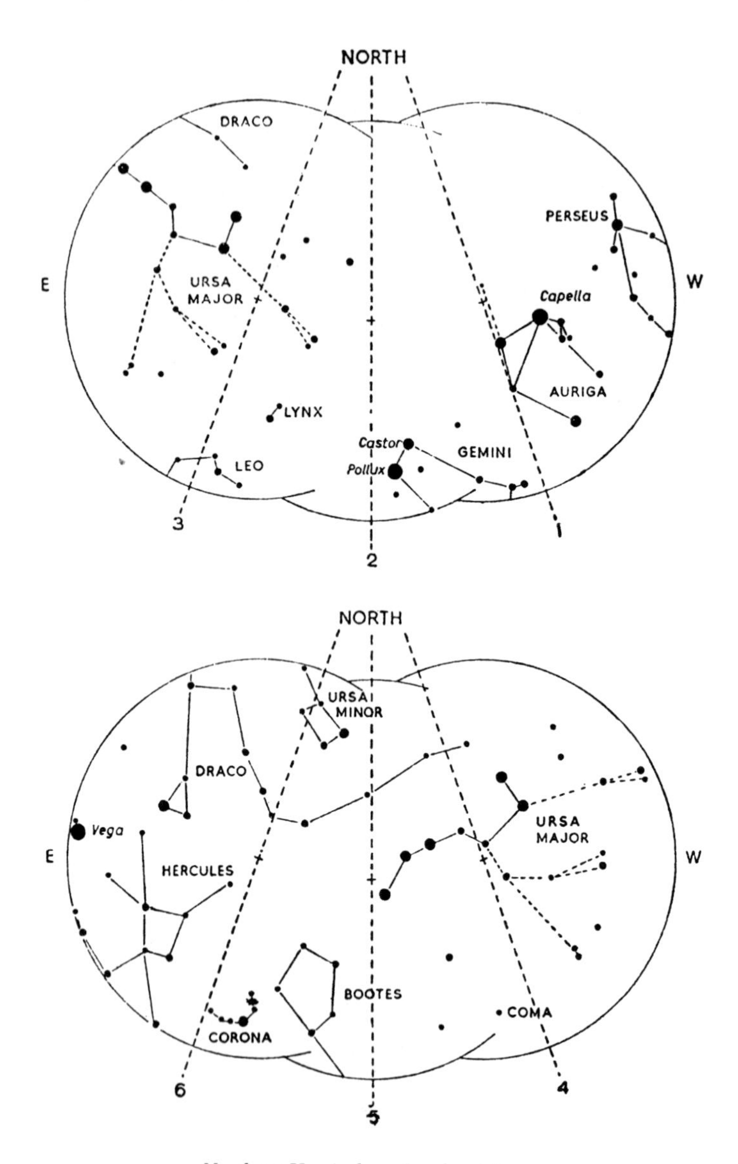

Northern Hemisphere Overhead Stars

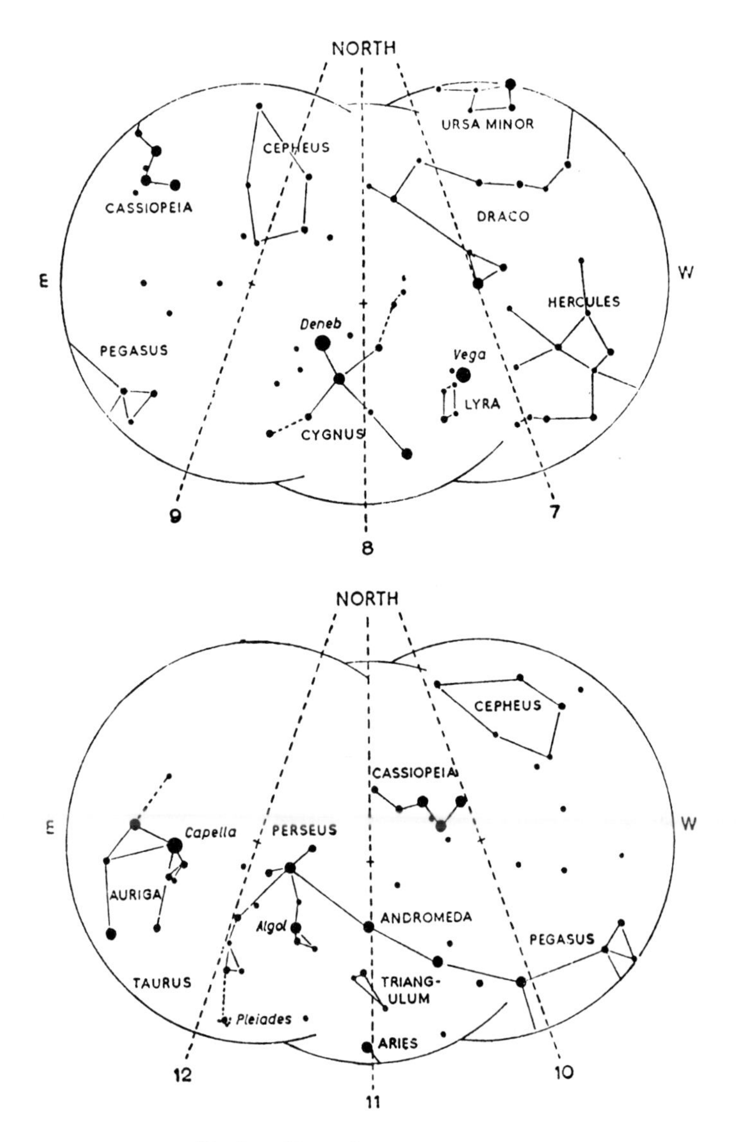

Northern Hemisphere Overhead Stars

1L

October 6 at	5h	October 21 at	4h
November 6 at	3h	November 21 at	2h
December 6 at	1h	December 21 at	midnight
January 6 at	23h	January 21 at	22h
February 6 at	21h	February 21 at	20h

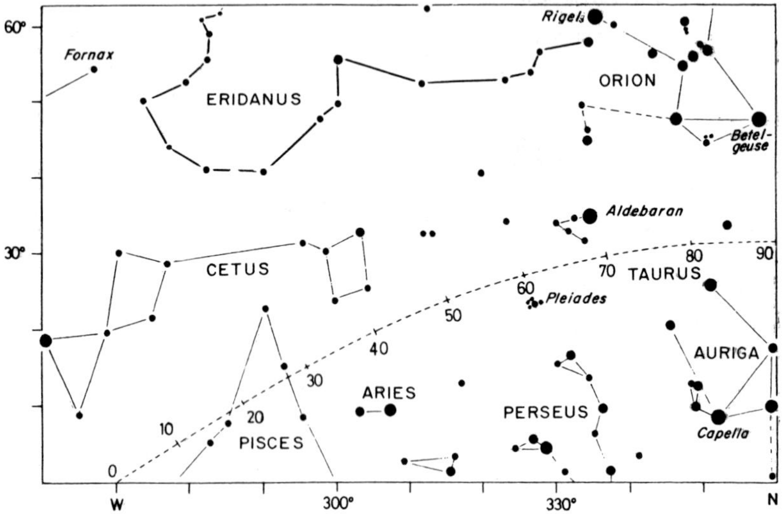

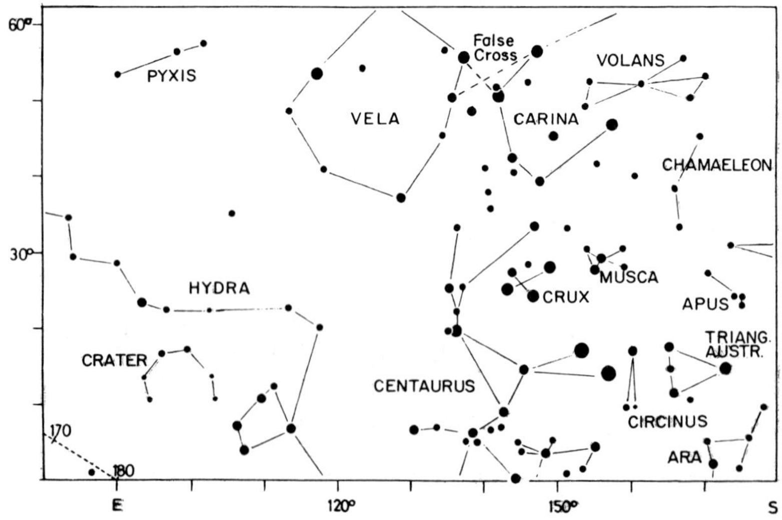

Fornax
ERIDANUS
Rigel
ORION
Betelgeuse
Aldebaran
CETUS
TAURUS
Pleiades
AURIGA
ARIES
PISCES
PERSEUS
Capella
0 10 20 30 40 50 60 70 80 90
60° 30°
W 300° 330° N

PYXIS
False Cross
VOLANS
VELA
CARINA
CHAMAELEON
HYDRA
MUSCA
CRUX
APUS
CRATER
TRIANG. AUSTR.
CENTAURUS
CIRCINUS
ARA
170 180
60° 30°
E 120° 150° S

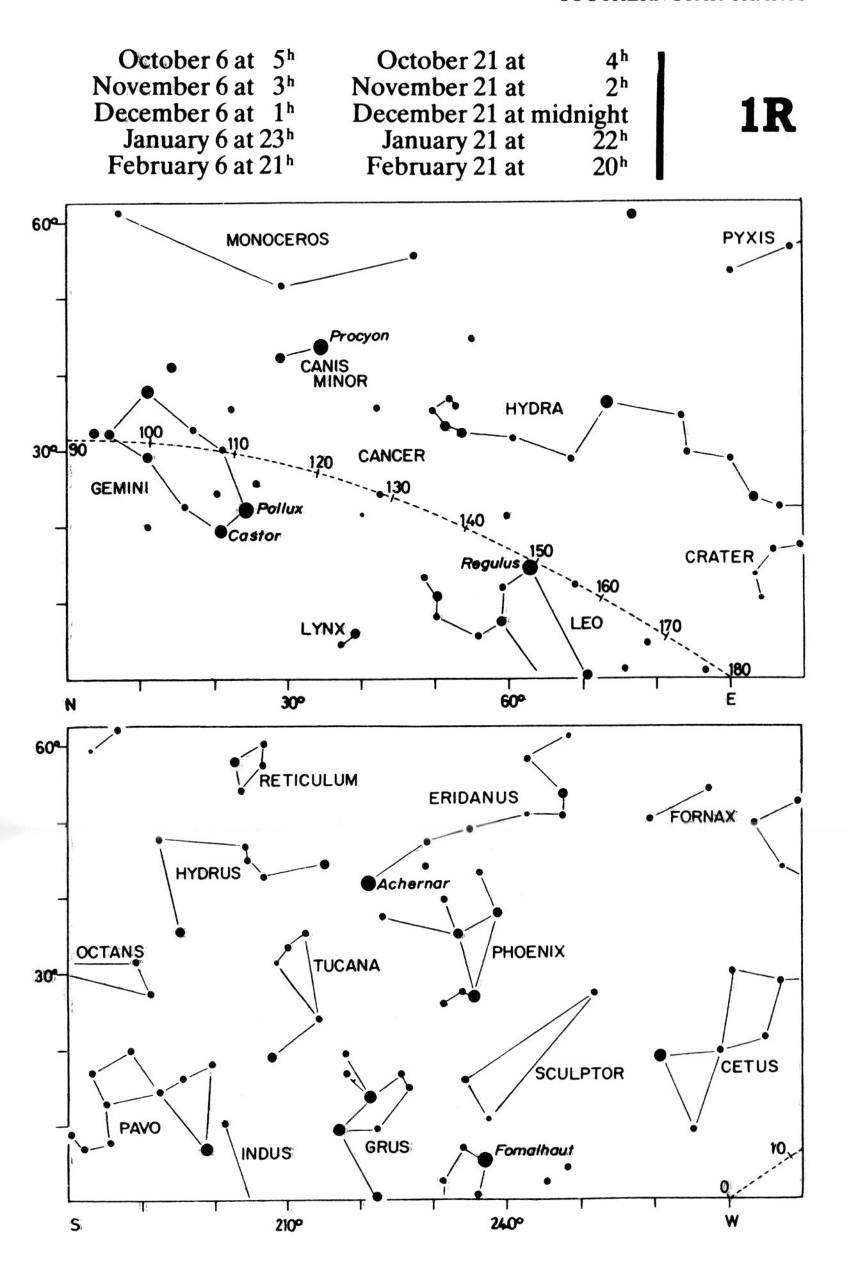
October 6 at 5h
November 6 at 3h
December 6 at 1h
January 6 at 23h
February 6 at 21h
October 21 at 4h
November 21 at 2h
December 21 at midnight
January 21 at 22h
February 21 at 20h
1R
60°
30°
MONOCEROS
PYXIS
Procyon
CANIS
MINOR
HYDRA
CANCER
GEMINI
Pollux
Castor
Regulus
CRATER
LEO
LYNX
90
100
110
120
130
140
150
160
170
180
N
30°
60°
E
RETICULUM
ERIDANUS
FORNAX
HYDRUS
Achernar
PHOENIX
OCTANS
TUCANA
SCULPTOR
CETUS
PAVO
INDUS
GRUS
Fomalhaut
10
0
S
210°
240°
W

2L

November 6 at 5ʰ	November 21 at 4ʰ
December 6 at 3ʰ	December 21 at 2ʰ
January 6 at 1ʰ	January 21 at midnight
February 6 at 23ʰ	February 21 at 22ʰ
March 6 at 21ʰ	March 21 at 20ʰ

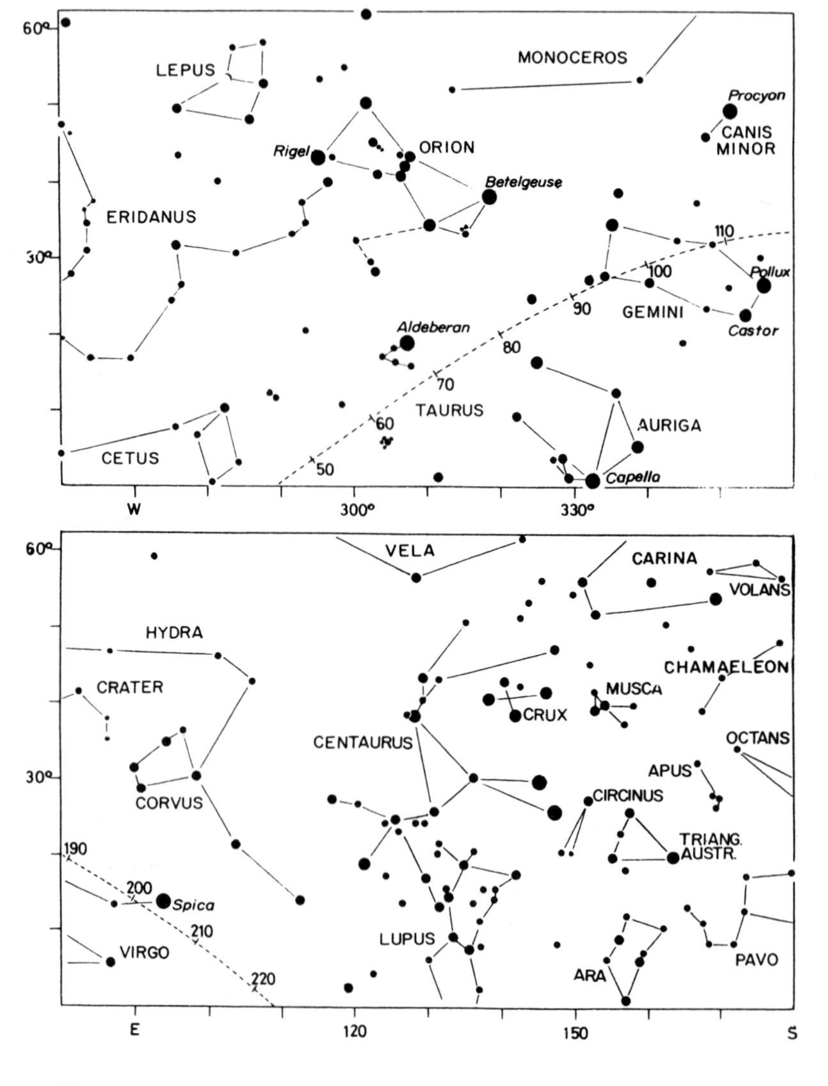

November 6 at 5h	November 21 at 4h
December 6 at 3h	December 21 at 2h
January 6 at 1h	January 21 at midnight
February 6 at 23h	February 21 at 22h
March 6 at 21h	March 21 at 20h

2R

3L

January 6 at 3^h	January 21 at 2^h
February 6 at 1^h	February 21 at midnight
March 6 at 23^h	March 21 at 22^h
April 6 at 21^h	April 21 at 20^h
May 6 at 19^h	May 21 at 18^h

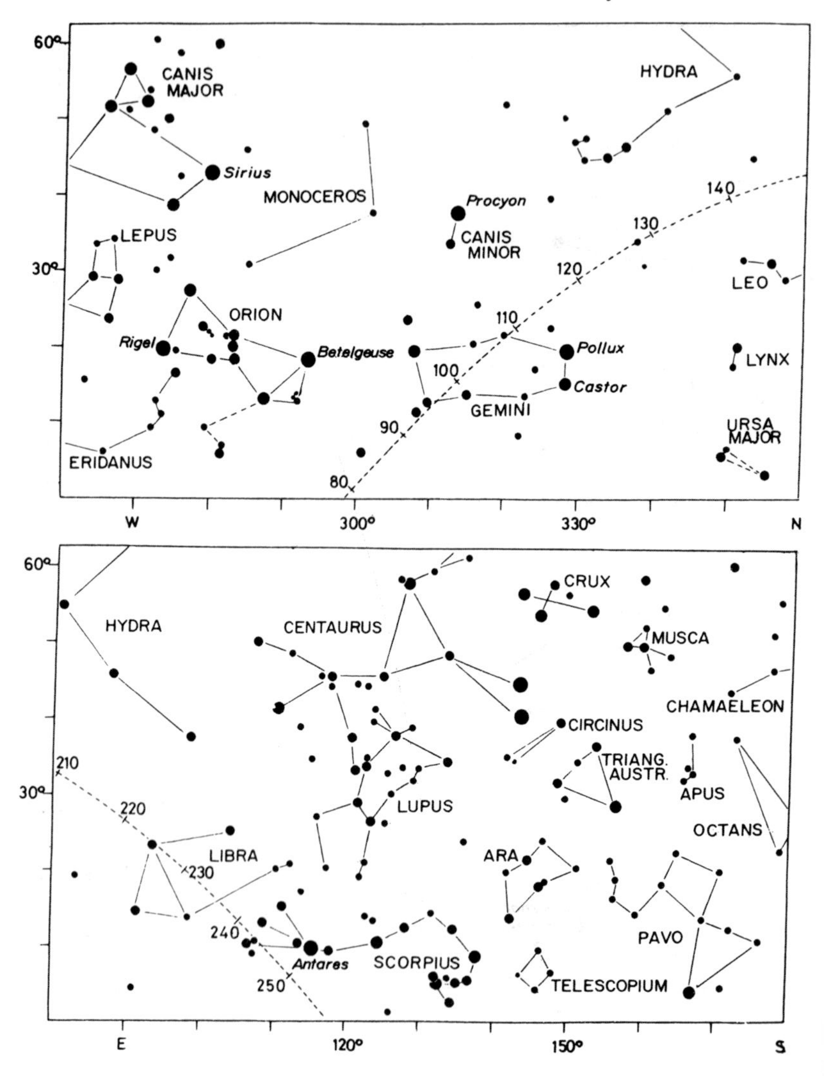

January 6 at 3^h — January 21 at 2^h
February 6 at 1^h — February 21 at midnight
March 6 at 23^h — March 21 at 22^h
April 6 at 21^h — April 21 at 20^h
May 6 at 19^h — May 21 at 18^h

3R

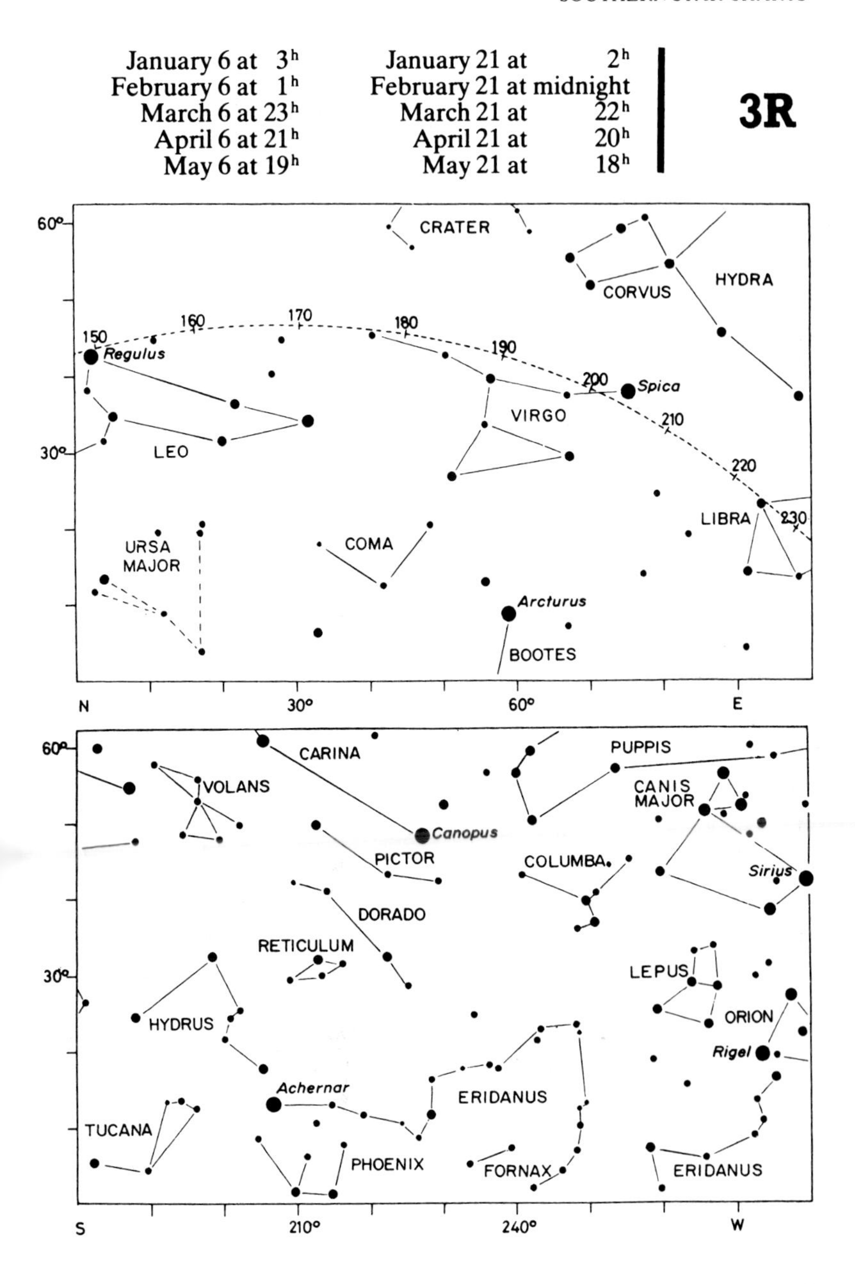

4L

February 6 at 3h	February 21 at 2h
March 6 at 1h	March 21 at midnight
April 6 at 23h	April 21 at 22h
May 6 at 21h	May 21 at 20h
June 6 at 19h	June 21 at 18h

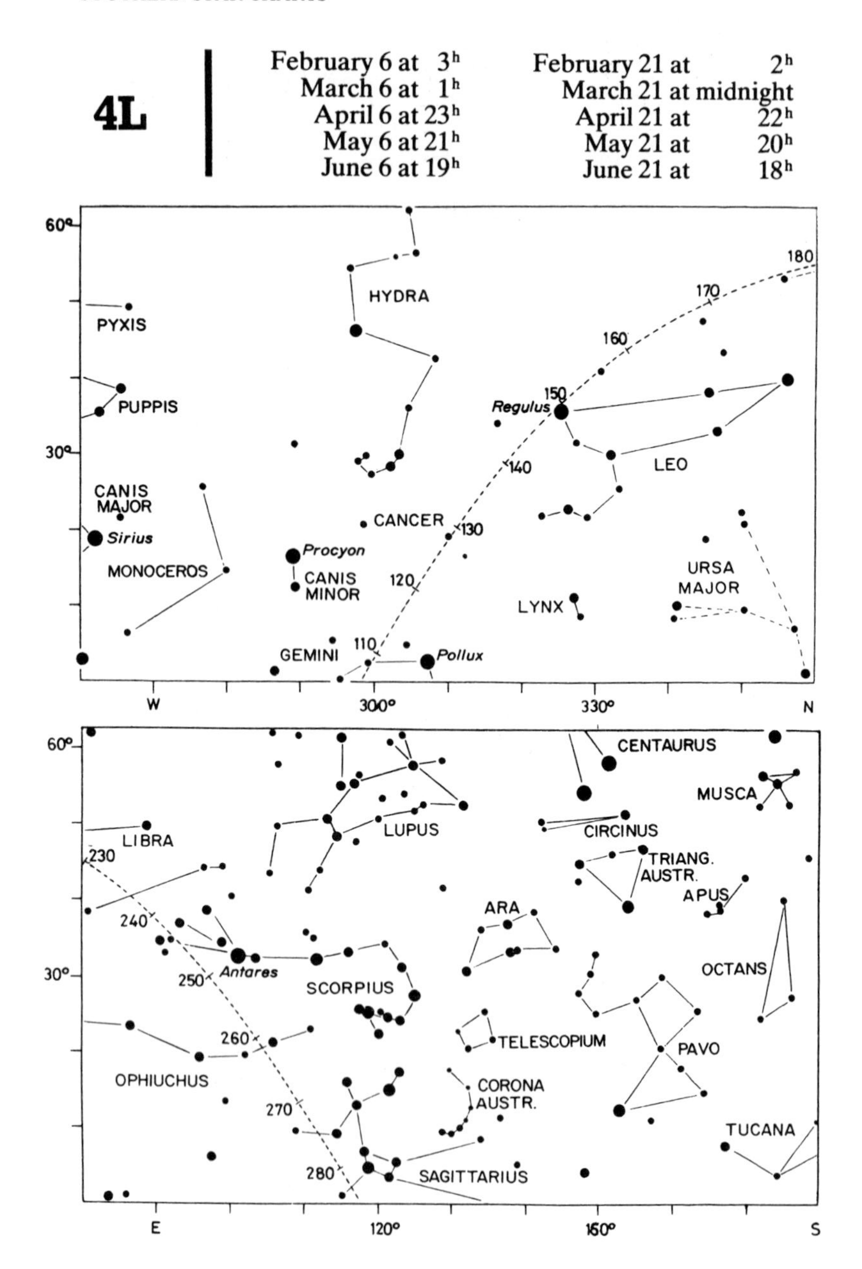

February 6 at 3h	February 21 at 2h	**4R**
March 6 at 1h	March 21 at midnight	
April 6 at 23h	April 21 at 22h	
May 6 at 21h	May 21 at 20h	
June 6 at 19h	June 21 at 18h	

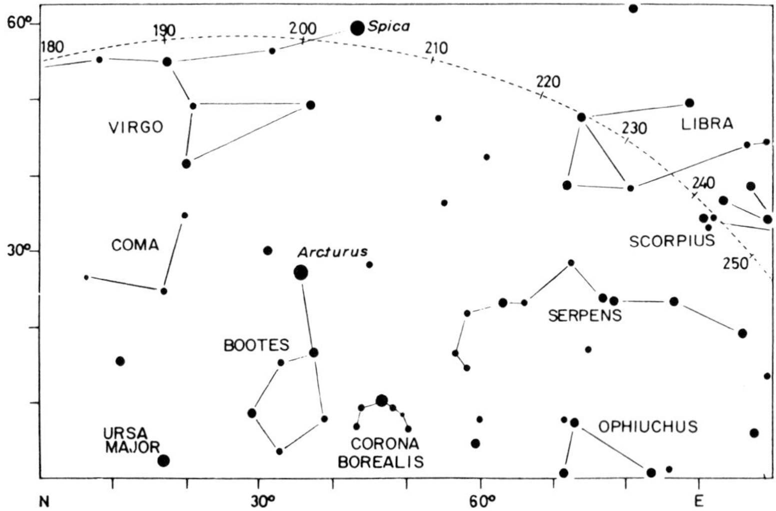

60° 180 190 200 Spica 210 220 230 240 250
VIRGO
LIBRA
COMA
30°
Arcturus
SCORPIUS
SERPENS
BOOTES
URSA MAJOR
CORONA BOREALIS
OPHIUCHUS
N 30° 60° E

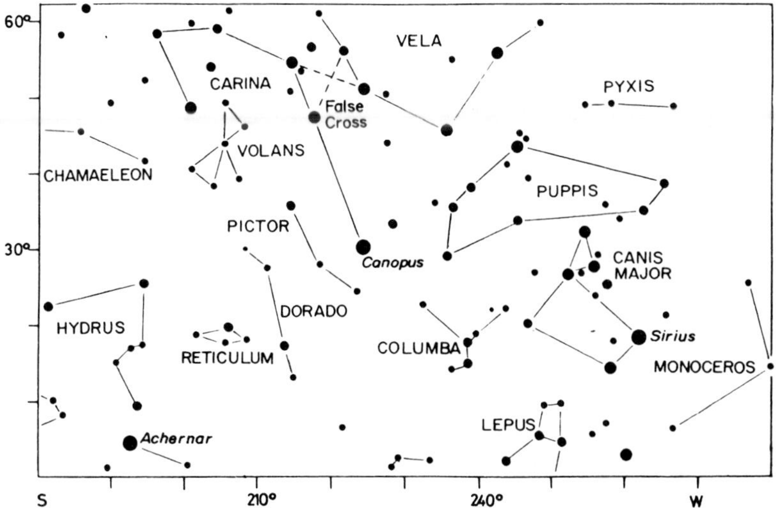

60°
VELA
CARINA
PYXIS
False Cross
VOLANS
CHAMAELEON
PUPPIS
PICTOR
30°
Canopus
CANIS MAJOR
DORADO
HYDRUS
Sirius
RETICULUM
COLUMBA
MONOCEROS
LEPUS
Achernar
S 210° 240° W

5L

March 6 at 3h	March 21 at 2h
April 6 at 1h	April 21 at midnight
May 6 at 23h	May 21 at 22h
June 6 at 21h	June 21 at 20h
July 6 at 19h	July 21 at 18h

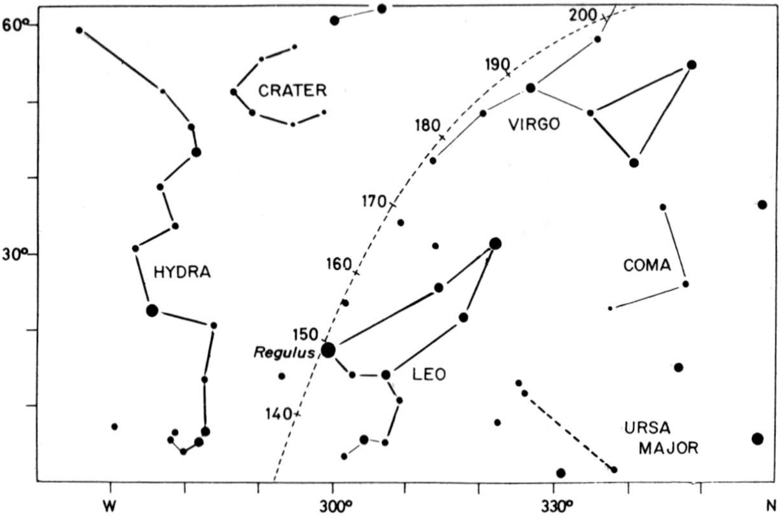

60°
30°
CRATER
HYDRA
VIRGO
COMA
LEO
Regulus
URSA MAJOR
200
190
180
170
160
150
140
W
300°
330°
N

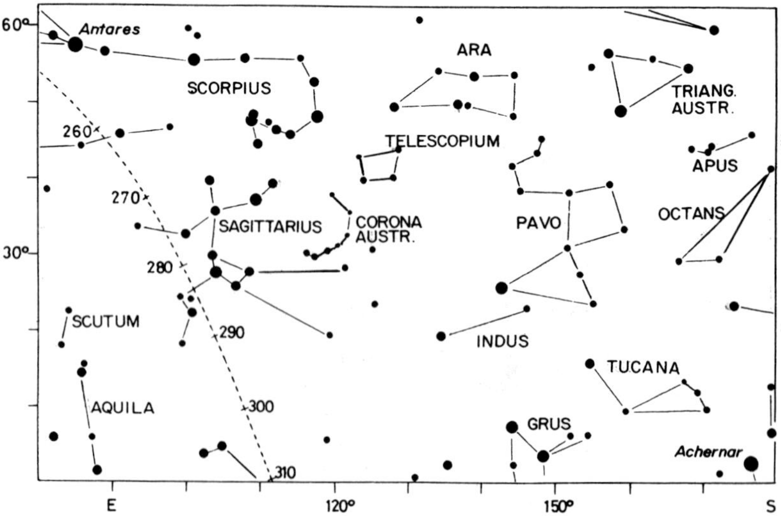

60°
30°
Antares
SCORPIUS
ARA
TRIANG. AUSTR.
TELESCOPIUM
APUS
OCTANS
SAGITTARIUS
CORONA AUSTR.
PAVO
SCUTUM
INDUS
TUCANA
AQUILA
GRUS
Achernar
260
270
280
290
300
310
E
120°
150°
S

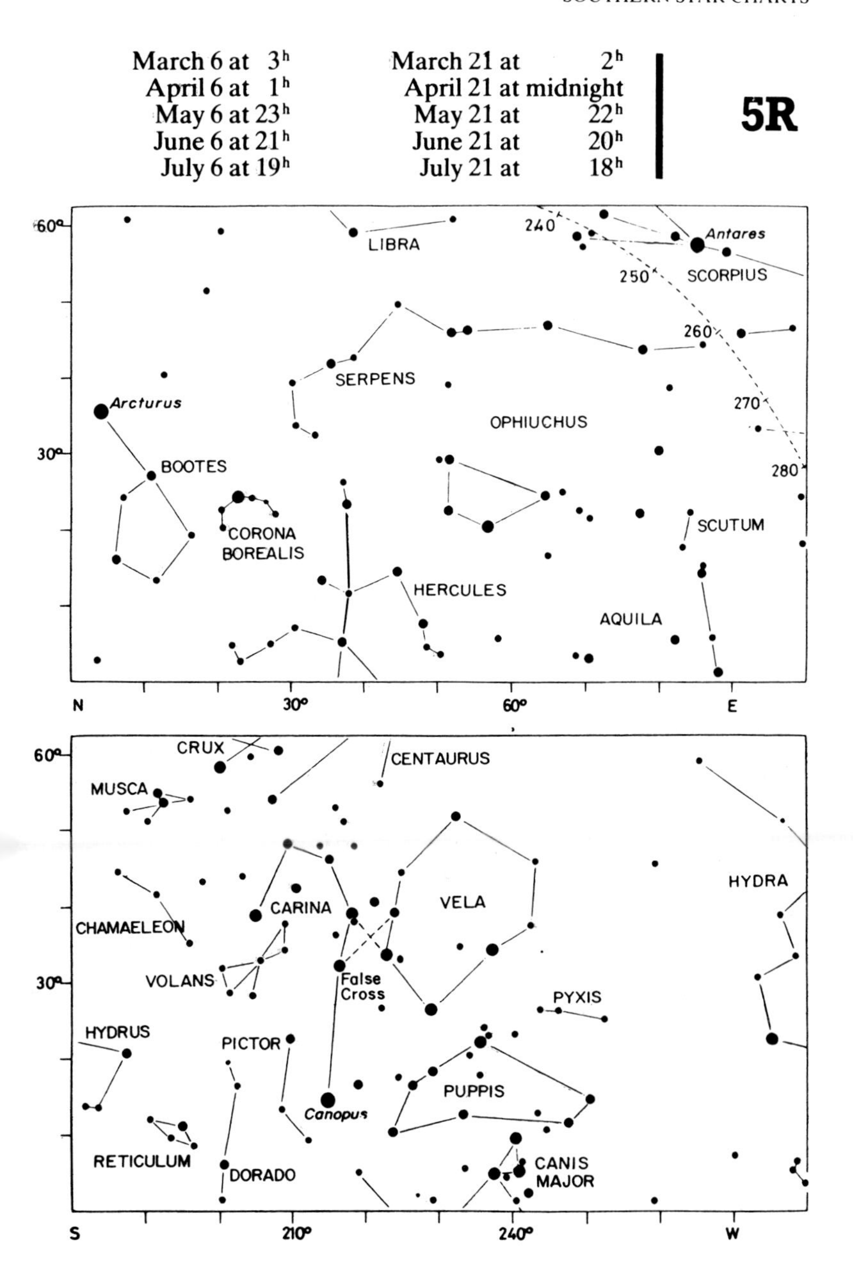
March 6 at 3h
April 6 at 1h
May 6 at 23h
June 6 at 21h
July 6 at 19h
March 21 at 2h
April 21 at midnight
May 21 at 22h
June 21 at 20h
July 21 at 18h
5R
60°
30°
LIBRA
240
Antares
250
SCORPIUS
260
SERPENS
270
Arcturus
OPHIUCHUS
280
BOOTES
CORONA
BOREALIS
SCUTUM
HERCULES
AQUILA
N
30°
60°
E
CRUX
CENTAURUS
MUSCA
HYDRA
CARINA
VELA
CHAMAELEON
VOLANS
False
Cross
PYXIS
HYDRUS
PICTOR
PUPPIS
Canopus
RETICULUM
DORADO
CANIS
MAJOR
S
210°
240°
W

6L

March 6 at 5ʰ	March 21 at 4ʰ
April 6 at 3ʰ	April 21 at 2ʰ
May 6 at 1ʰ	May 21 at midnight
June 6 at 23ʰ	June 21 at 22ʰ
July 6 at 21ʰ	July 21 at 20ʰ

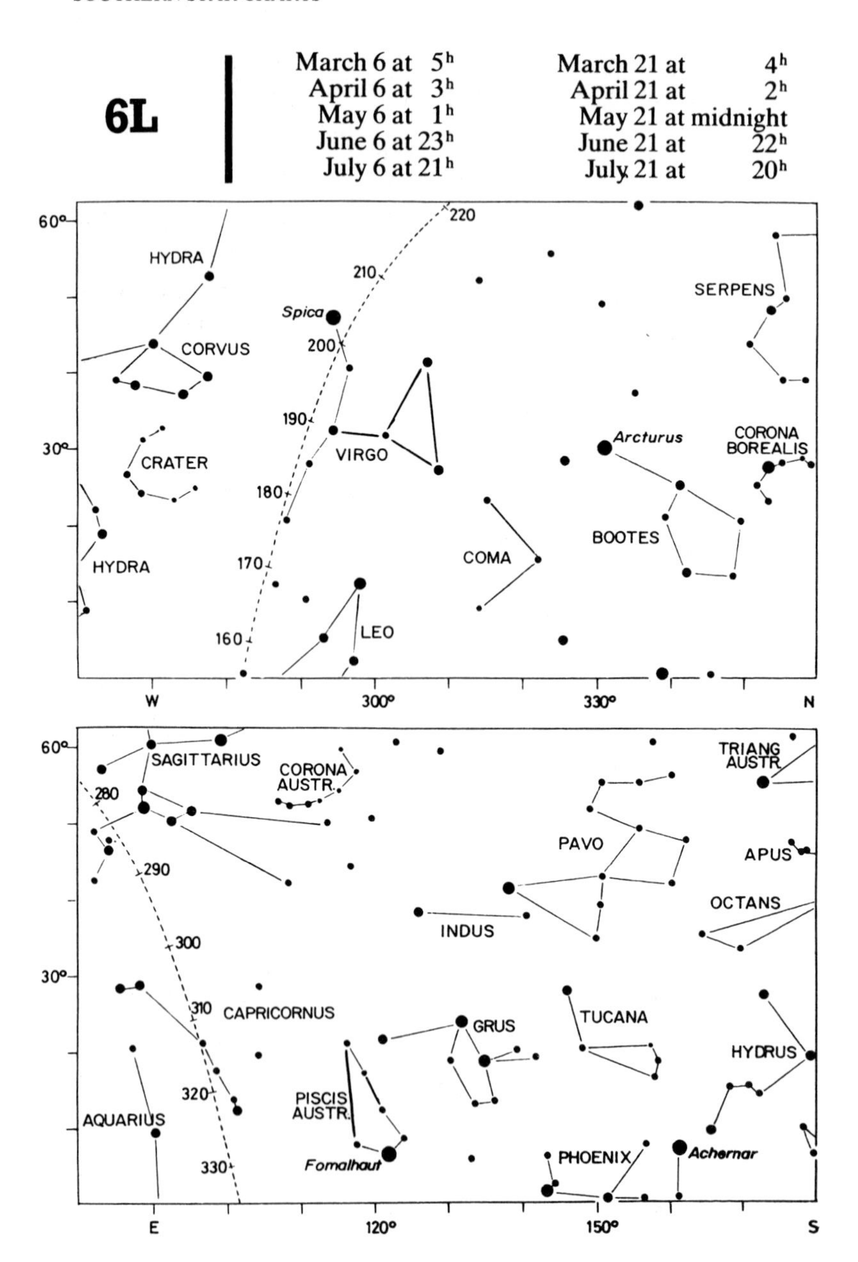

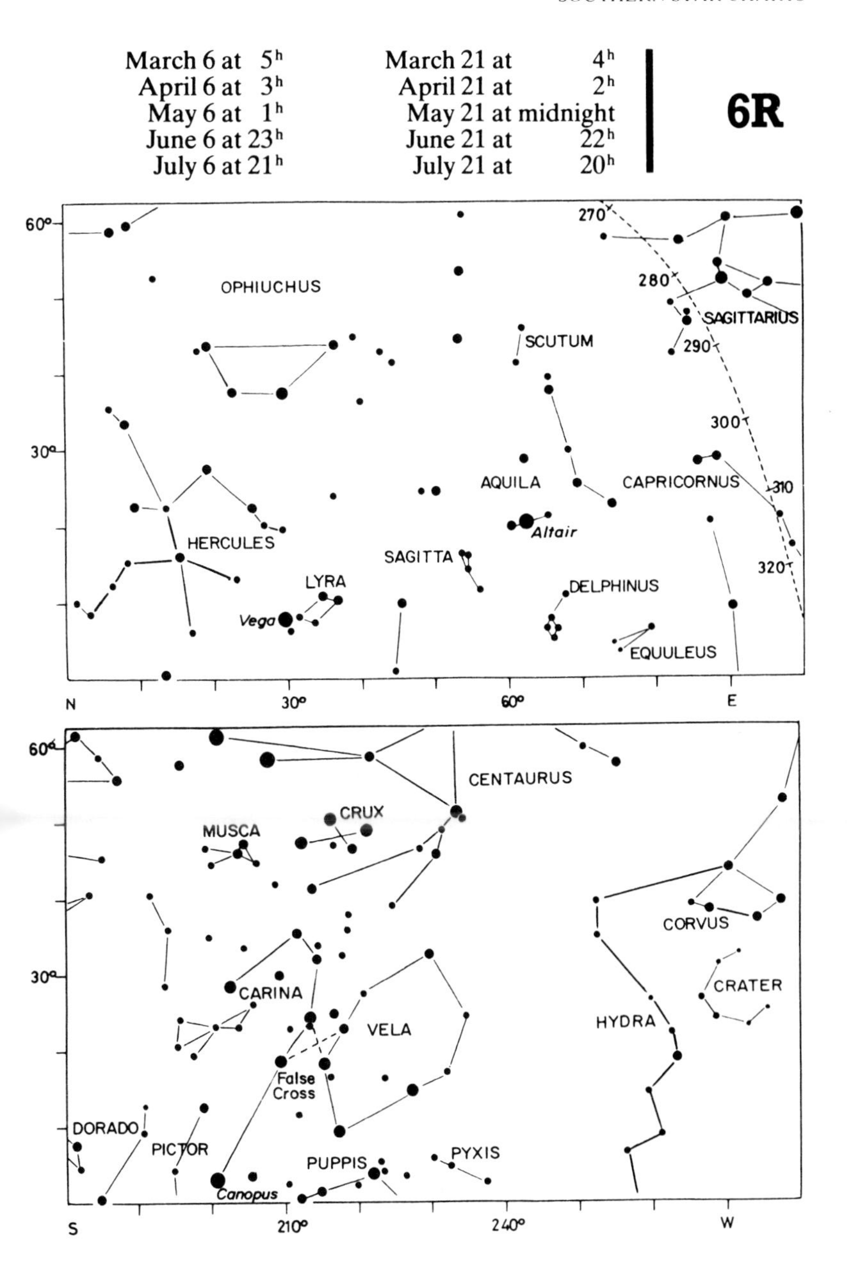
March 6 at 5h
April 6 at 3h
May 6 at 1h
June 6 at 23h
July 6 at 21h
March 21 at 4h
April 21 at 2h
May 21 at midnight
June 21 at 22h
July 21 at 20h
6R
60°
30°
OPHIUCHUS
270
280
290
300
310
320
SAGITTARIUS
SCUTUM
AQUILA
CAPRICORNUS
Altair
HERCULES
SAGITTA
LYRA
Vega
DELPHINUS
EQUULEUS
N
30°
60°
E
60°
30°
CENTAURUS
CRUX
MUSCA
CORVUS
CARINA
CRATER
VELA
HYDRA
False Cross
DORADO
PICTOR
PUPPIS
PYXIS
Canopus
S
210°
240°
W

7L

April 6 at 5ʰ	April 21 at 4ʰ
May 6 at 3ʰ	May 21 at 2ʰ
June 6 at 1ʰ	June 21 at midnight
July 6 at 23ʰ	July 21 at 22ʰ
August 6 at 21ʰ	August 21 at 20ʰ

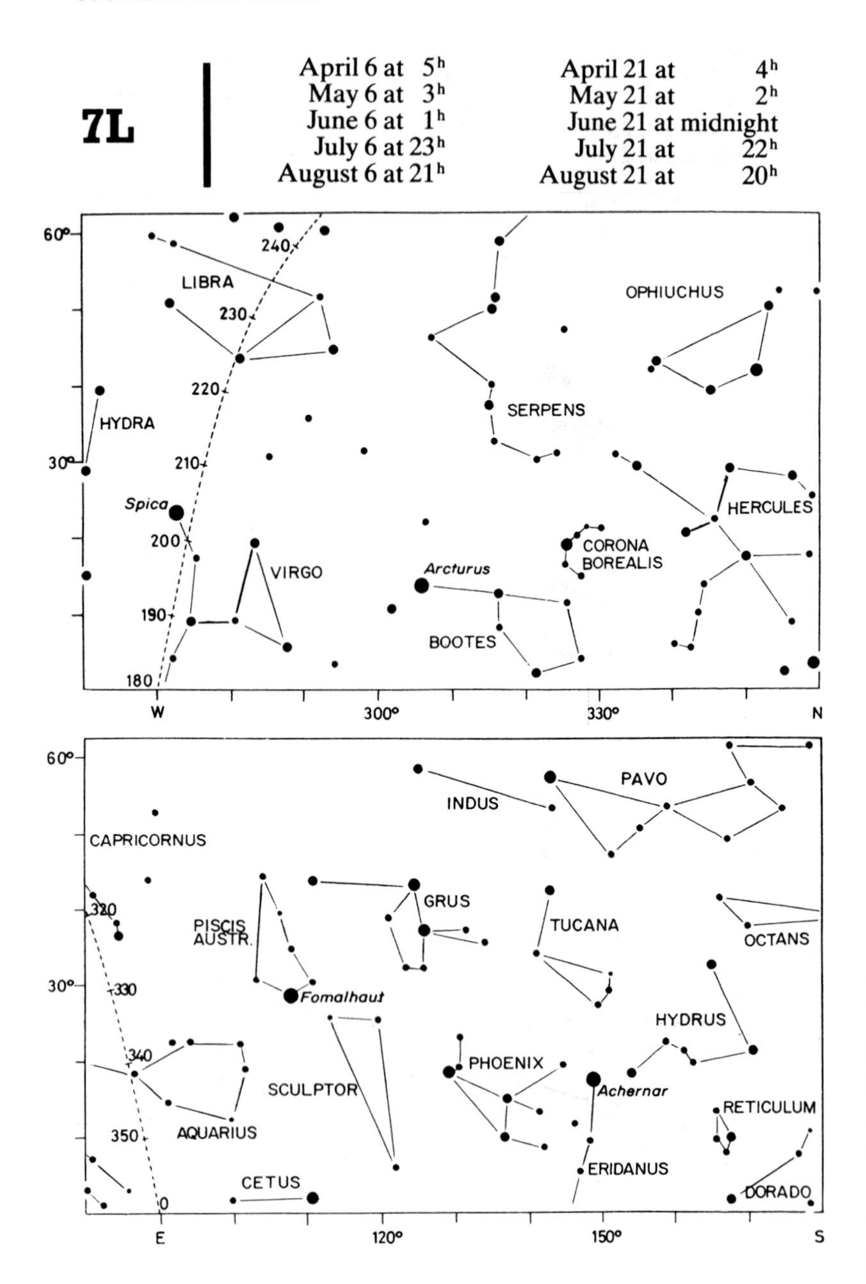

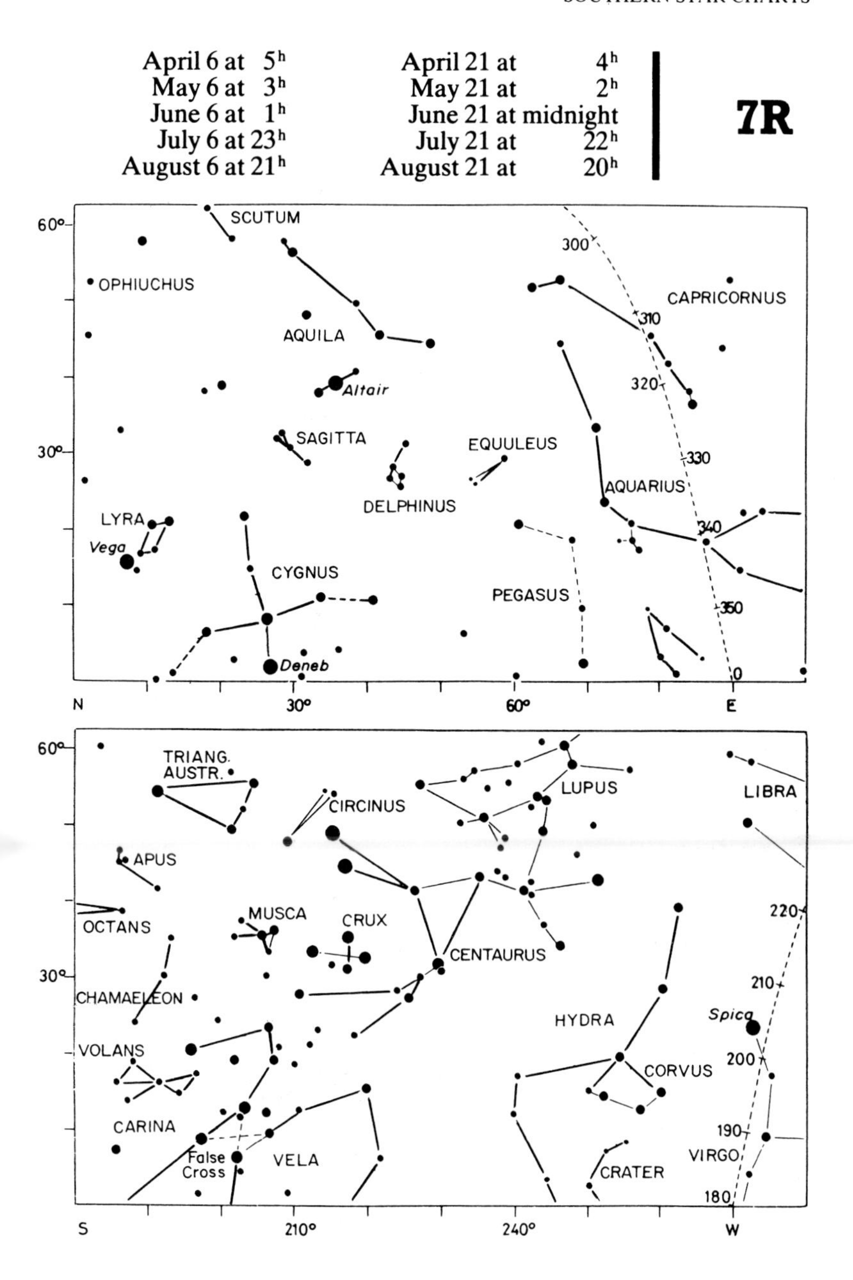
April 6 at 5^h
May 6 at 3^h
June 6 at 1^h
July 6 at 23^h
August 6 at 21^h
April 21 at 4^h
May 21 at 2^h
June 21 at midnight
July 21 at 22^h
August 21 at 20^h
7R
60°
30°
SCUTUM
OPHIUCHUS
AQUILA
Altair
SAGITTA
DELPHINUS
EQUULEUS
CAPRICORNUS
AQUARIUS
PEGASUS
LYRA
Vega
CYGNUS
Deneb
300
310
320
330
340
350
0
N
30°
60°
E
60°
30°
TRIANG. AUSTR.
CIRCINUS
LUPUS
LIBRA
APUS
OCTANS
MUSCA
CRUX
CENTAURUS
CHAMAELEON
HYDRA
Spica
VOLANS
CORVUS
CARINA
False Cross
VELA
CRATER
VIRGO
220
210
200
190
180
S
210°
240°
W

8L

May 6 at 5h	May 21 at 4h
June 6 at 3h	June 21 at 2h
July 6 at 1h	July 21 at midnight
August 6 at 23h	August 21 at 22h
September 6 at 21h	September 21 at 20h

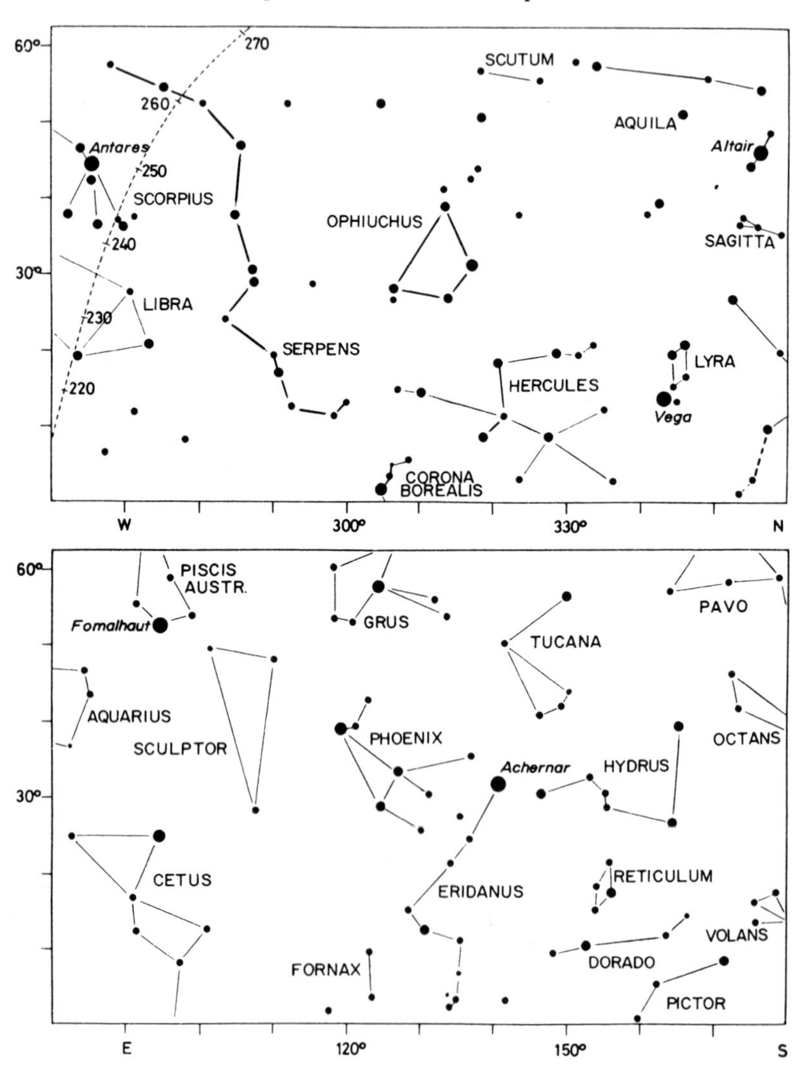

May 6 at 5^h	May 21 at 4^h	**8R**
June 6 at 3^h	June 21 at 2^h	
July 6 at 1^h	July 21 at midnight	
August 6 at 23^h	August 21 at 22^h	
September 6 at 21^h	September 21 at 20^h	

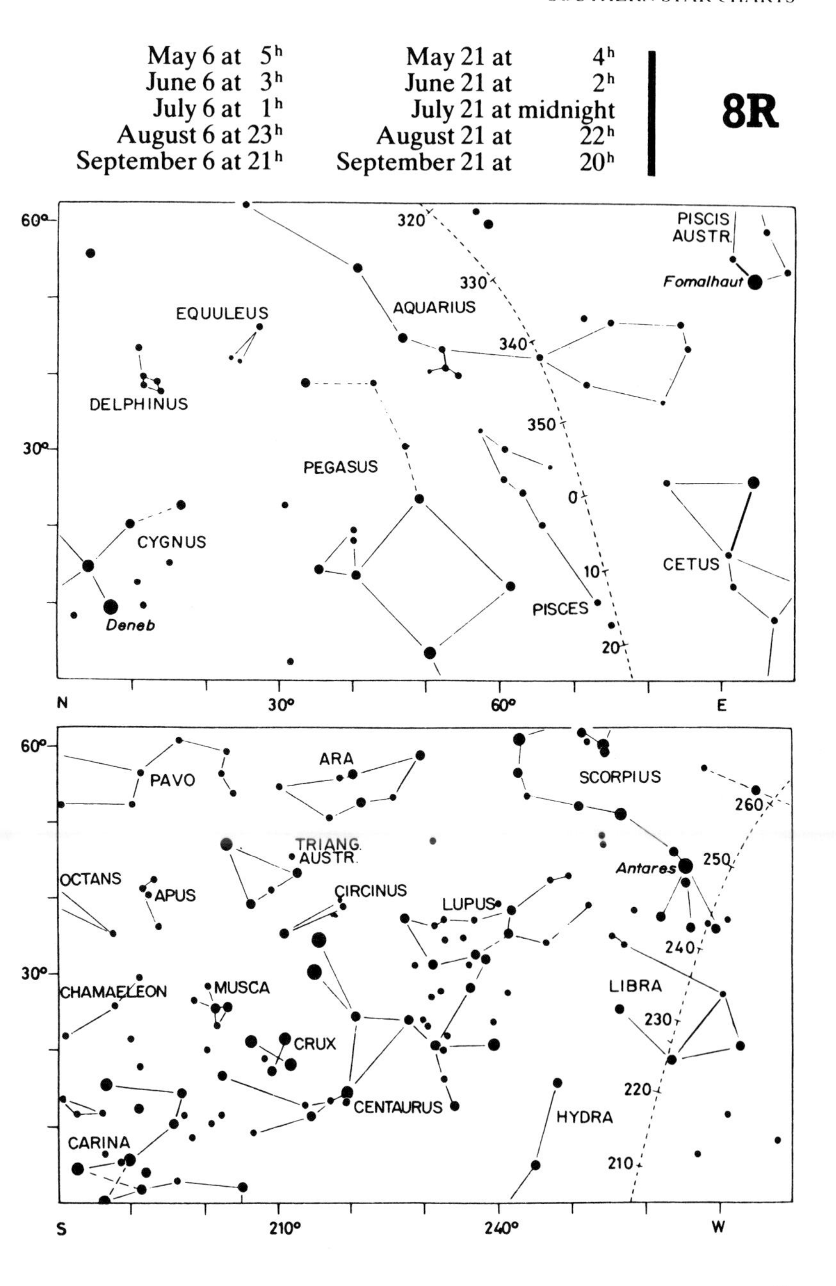

9L

June 6 at 5h	June 21 at 4h
July 6 at 3h	July 21 at 2h
August 6 at 1h	August 21 at midnight
September 6 at 23h	September 21 at 22h
October 6 at 21h	October 21 at 20h

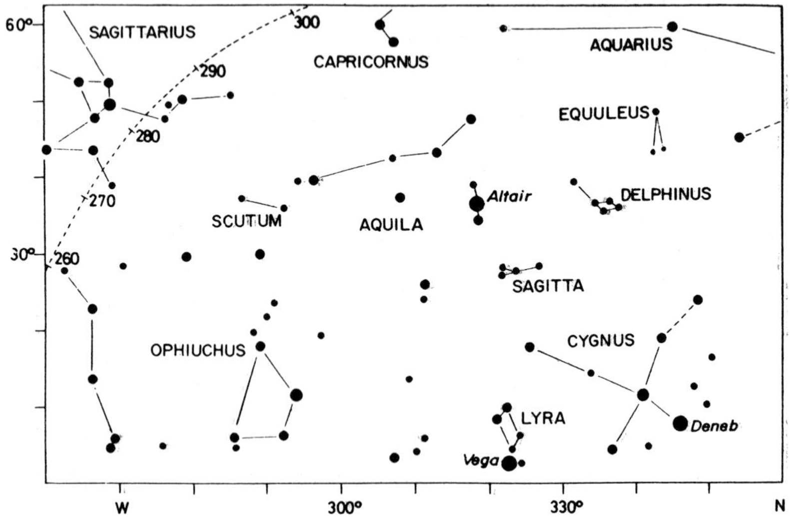

60°
SAGITTARIUS
300
CAPRICORNUS
AQUARIUS
290
EQUULEUS
280
270
SCUTUM
AQUILA
Altair
DELPHINUS
30°
260
SAGITTA
OPHIUCHUS
CYGNUS
LYRA
Deneb
Vega
W
300°
330°
N

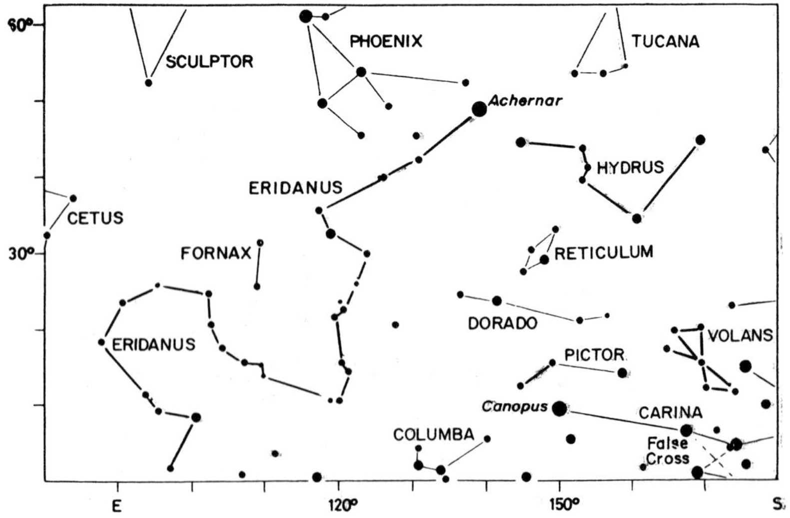

60°
SCULPTOR
PHOENIX
TUCANA
Achernar
HYDRUS
ERIDANUS
CETUS
FORNAX
30°
RETICULUM
DORADO
ERIDANUS
VOLANS
PICTOR
Canopus
CARINA
COLUMBA
False
Cross
E
120°
150°
S

June 6 at 5h	June 21 at 4h
July 6 at 3h	July 21 at 2h
August 6 at 1h	August 21 at midnight
September 6 at 23h	September 21 at 22h
October 6 at 21h	October 21 at 20h

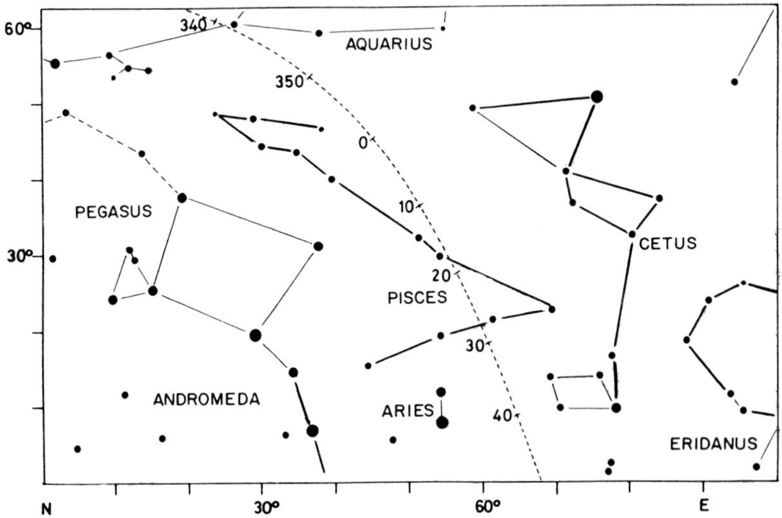

9R

60°
340
AQUARIUS
350
0
PEGASUS
10
CETUS
30°
20
PISCES
30
ANDROMEDA
ARIES
40
ERIDANUS
N
30°
60°
E

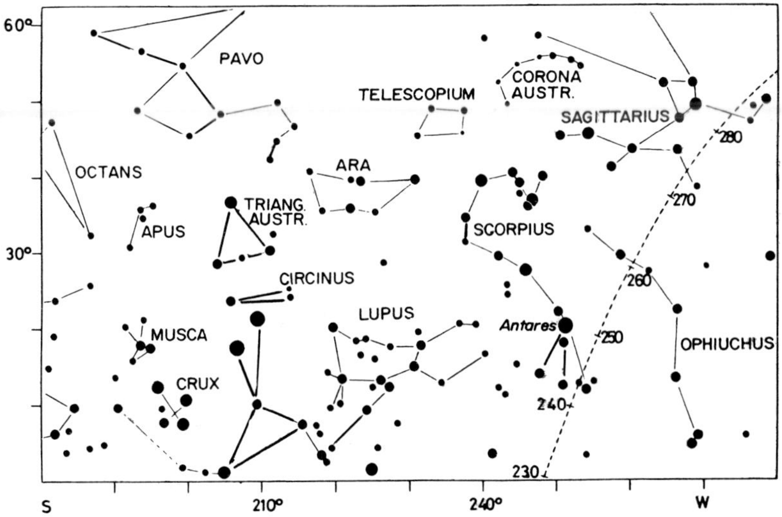

60°
PAVO
CORONA AUSTR.
TELESCOPIUM
SAGITTARIUS
280
OCTANS
ARA
270
TRIANG. AUSTR.
APUS
SCORPIUS
30°
CIRCINUS
260
LUPUS
Antares
250
MUSCA
OPHIUCHUS
CRUX
240
230
S
210°
240°
W

10L

July 6 at 5ʰ	July 21 at 4ʰ
August 6 at 3ʰ	August 21 at 2ʰ
September 6 at 1ʰ	September 21 at midnight
October 6 at 23ʰ	October 21 at 22ʰ
November 6 at 21ʰ	November 21 at 20ʰ

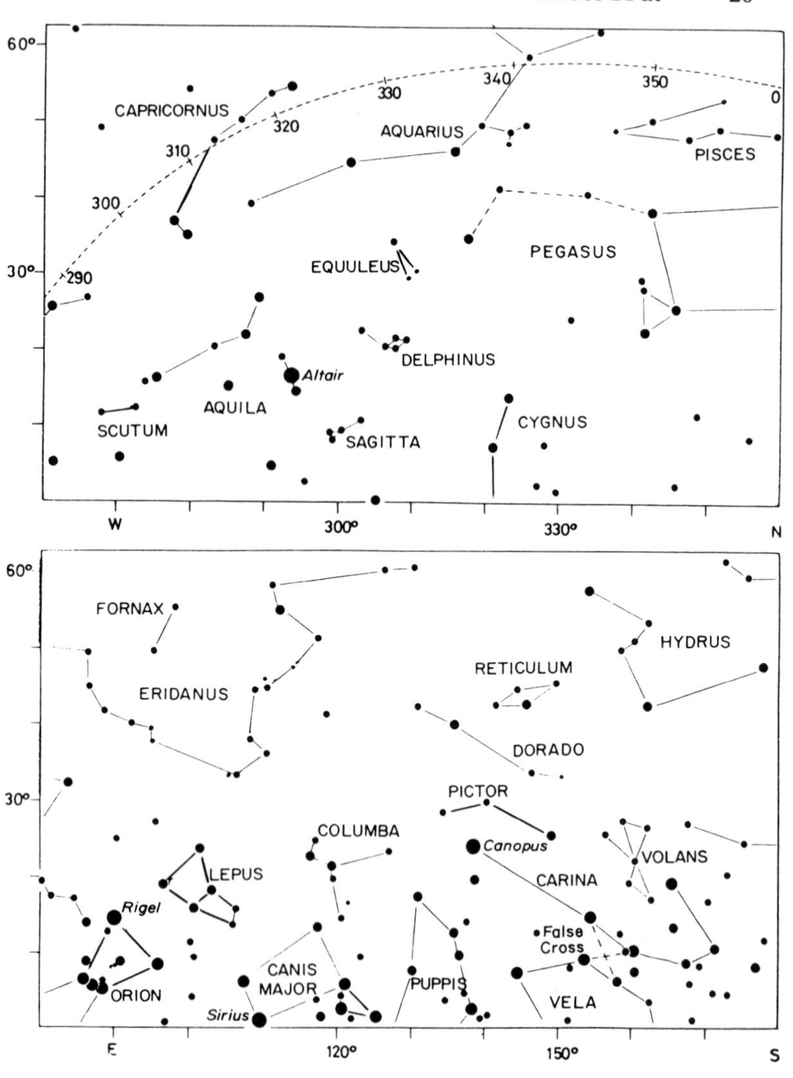

July 6 at 5ʰ	July 21 at 4ʰ	**10R**
August 6 at 3ʰ	August 21 at 2ʰ	
September 6 at 1ʰ	September 21 at midnight	
October 6 at 23ʰ	October 21 at 22ʰ	
November 6 at 21ʰ	November 21 at 20ʰ	

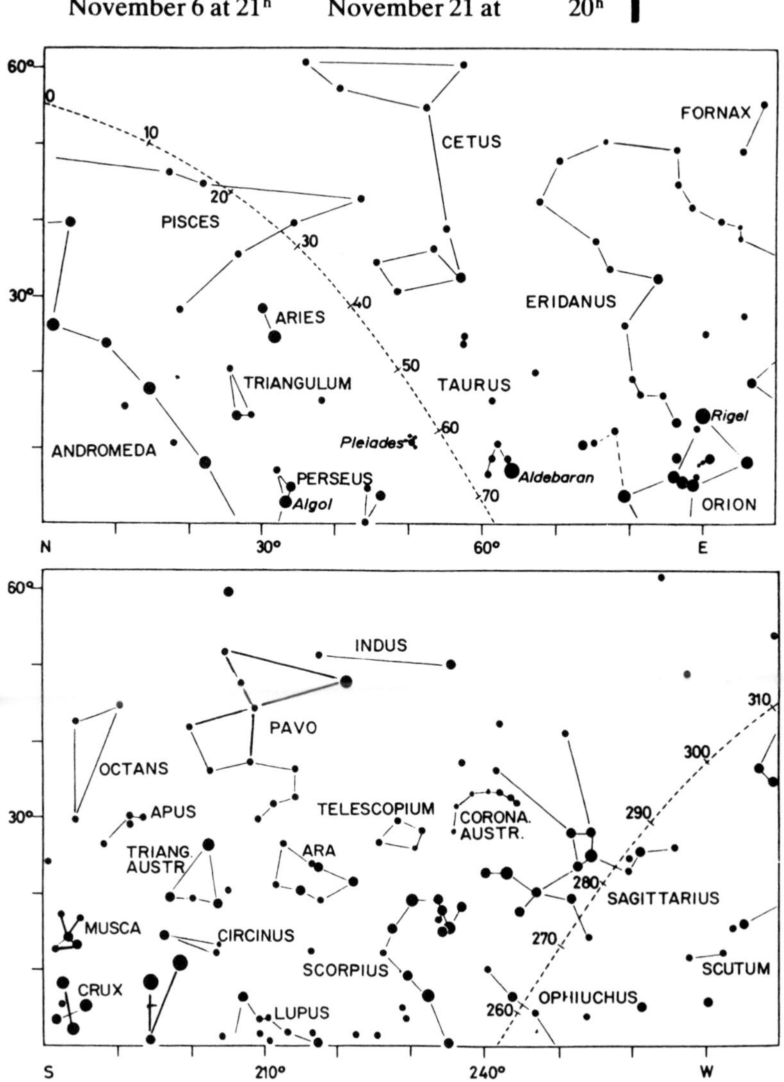

11L

August 6 at 5h	August 21 at 4h
September 6 at 3h	September 21 at 2h
October 6 at 1h	October 21 at midnight
November 6 at 23h	November 21 at 22h
December 6 at 21h	December 21 at 20h

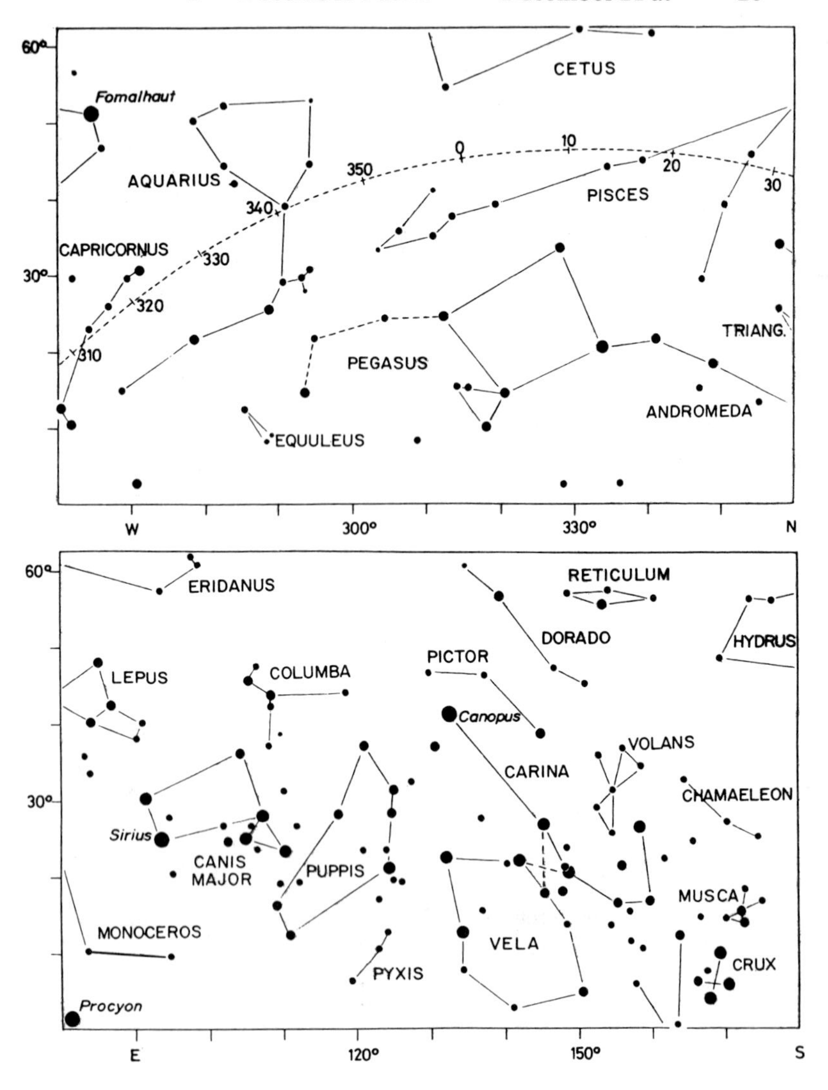

August 6 at 5h	August 21 at 4h	**11R**
September 6 at 3h	September 21 at 2h	
October 6 at 1h	October 21 at midnight	
November 6 at 23h	November 21 at 22h	
December 6 at 21h	December 21 at 20h	

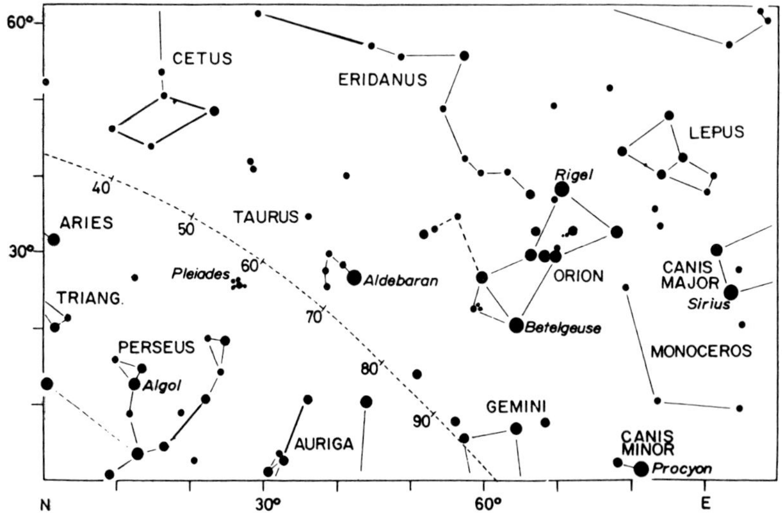

60°
CETUS
ERIDANUS
LEPUS
Rigel
40
ARIES
50
TAURUS
30°
Pleiades
60
Aldebaran
ORION
CANIS MAJOR
Sirius
TRIANG.
70
Betelgeuse
PERSEUS
MONOCEROS
80
Algol
GEMINI
90
AURIGA
CANIS MINOR
Procyon
N
30°
60°
E

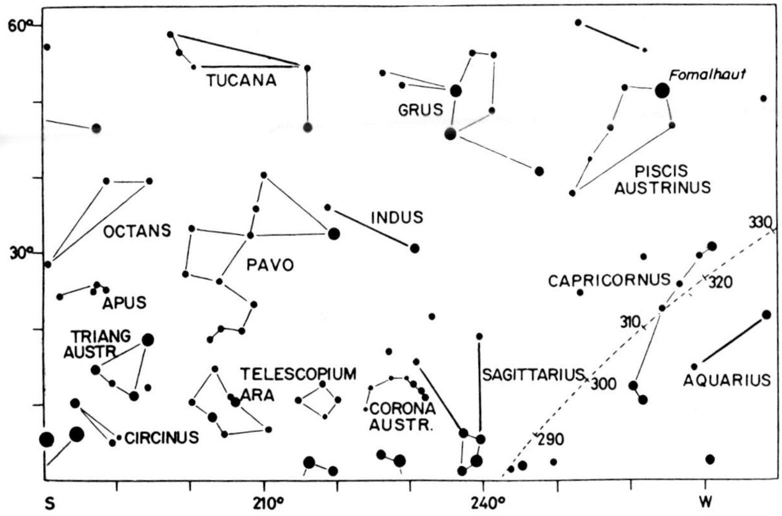

60°
TUCANA
Fomalhaut
GRUS
PISCIS AUSTRINUS
OCTANS
INDUS
330
30°
PAVO
CAPRICORNUS
320
APUS
310
TRIANG AUSTR
TELESCOPIUM
SAGITTARIUS
300
AQUARIUS
ARA
CORONA AUSTR.
CIRCINUS
290
S
210°
240°
W

12L

September 6 at 5^h	September 21 at 4^h
October 6 at 3^h	October 21 at 2^h
November 6 at 1^h	November 21 at midnight
December 6 at 23^h	December 21 at 22^h
January 6 at 21^h	January 21 at 20^h

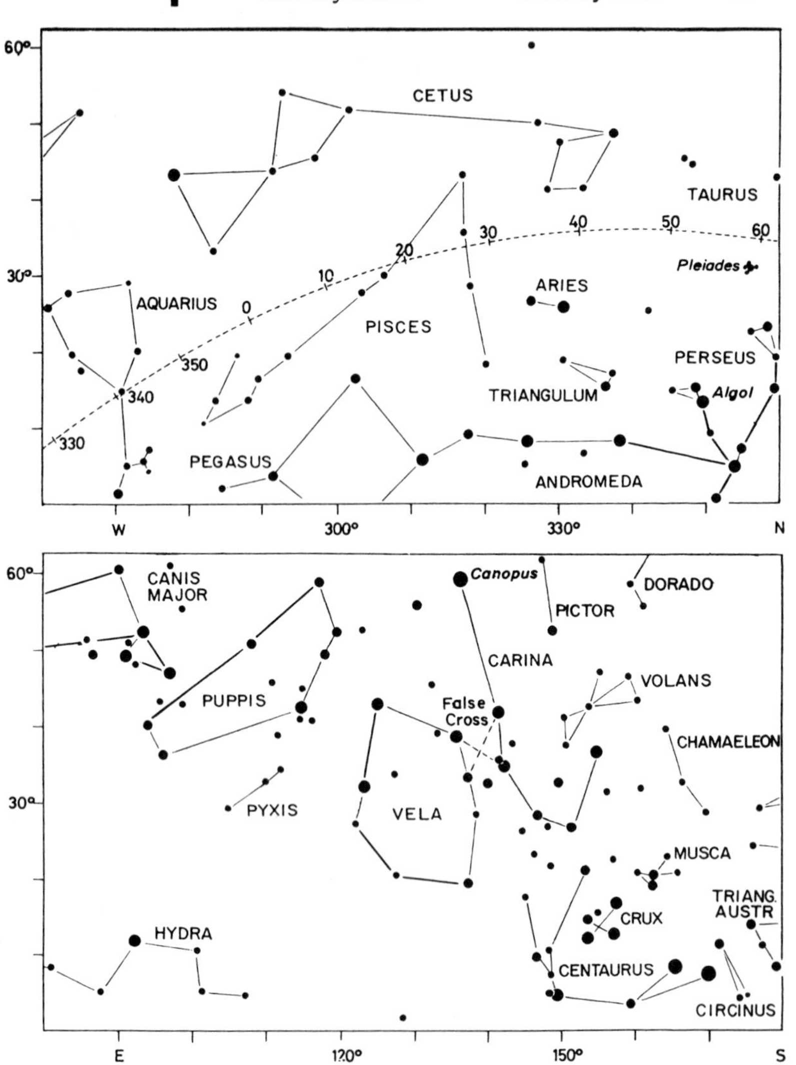

September 6 at 5h	September 21 at 4h	**12R**
October 6 at 3h	October 21 at 2h	
November 6 at 1h	November 21 at midnight	
December 6 at 23h	December 21 at 22h	
January 6 at 21h	January 21 at 20h	

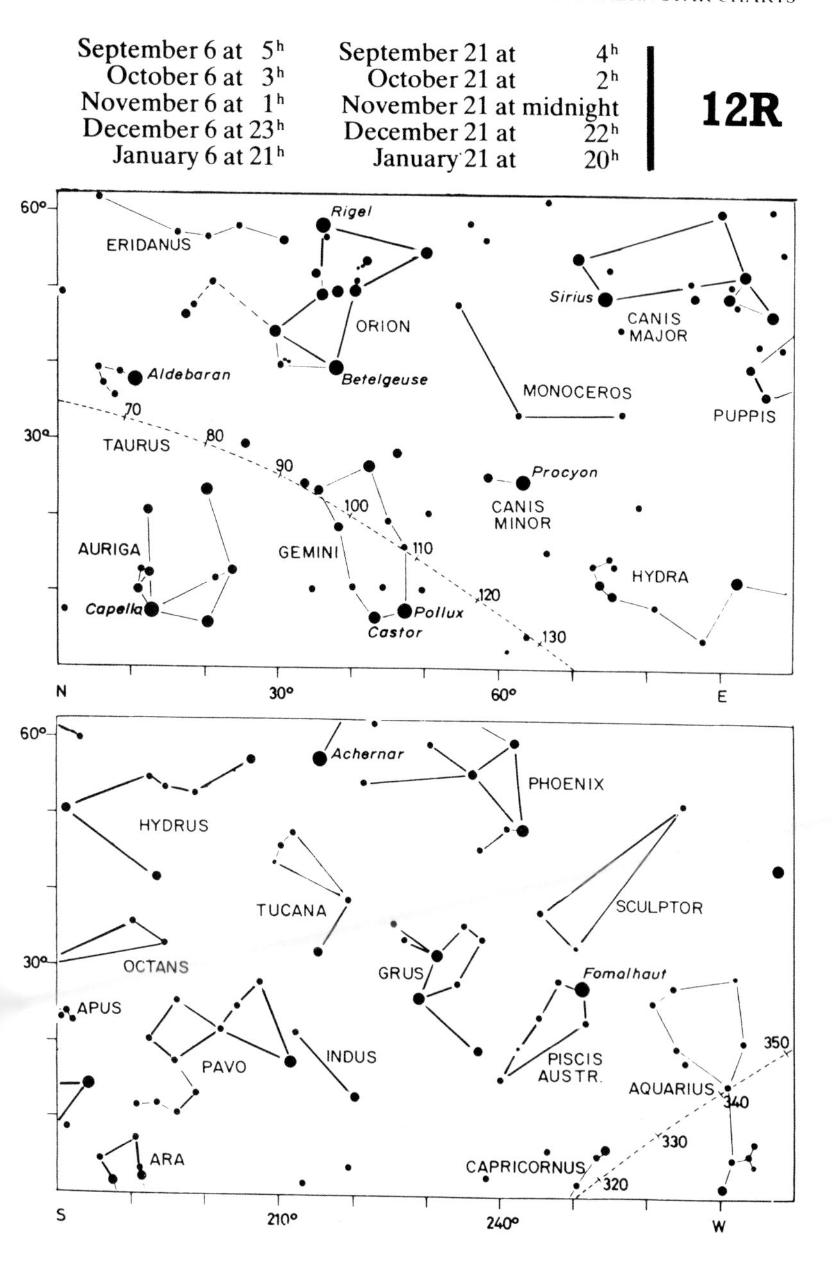

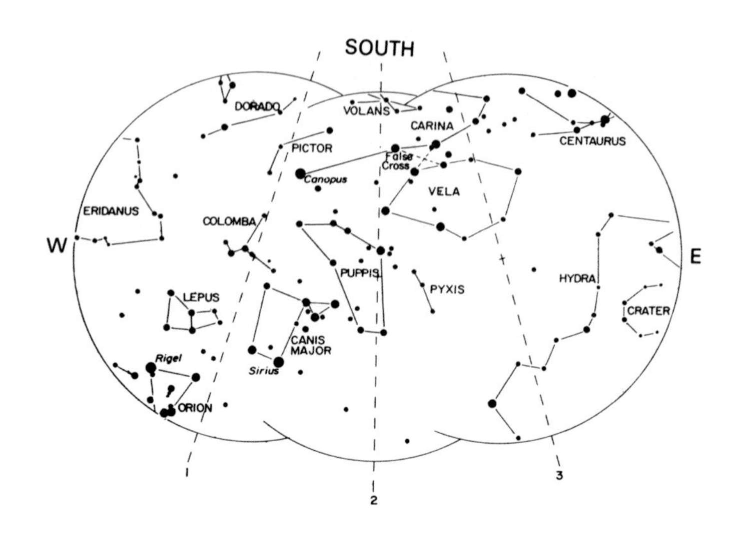

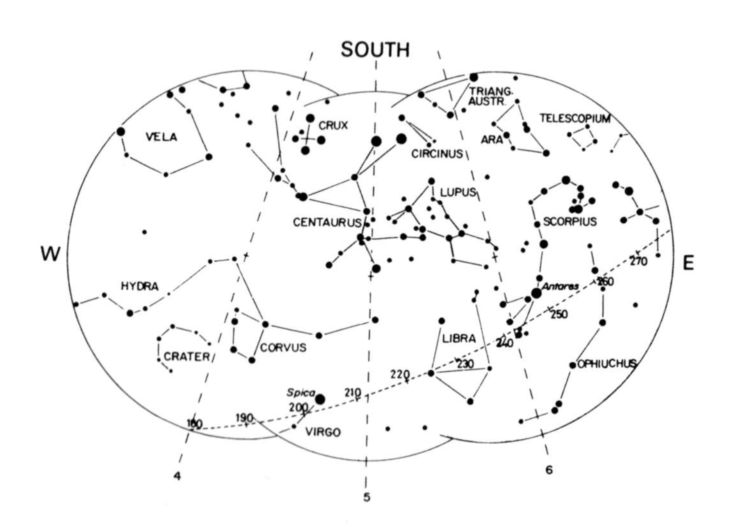

Southern Hemisphere Overhead Stars

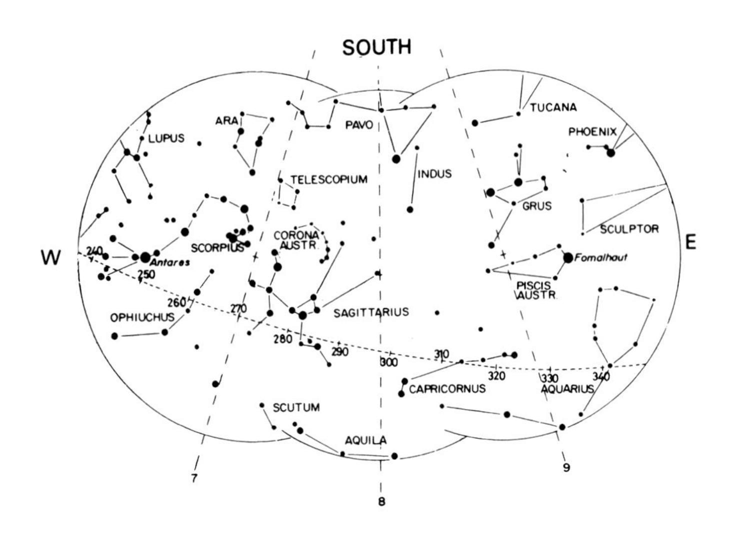

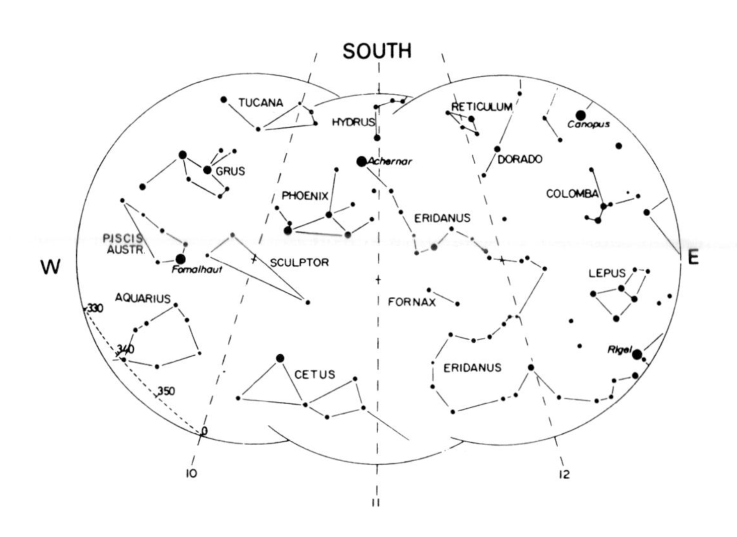

Southern Hemisphere Overhead Stars

The Planets and the Ecliptic

The paths of the planets about the Sun all lie close to the plane of the ecliptic, which is marked for us in the sky by the apparent path of the Sun among the stars, and is shown on the star charts by a broken line. The Moon and planets will always be found close to this line, never departing from it by more than about 7 degrees. Thus the planets are most favourably placed for observation when the ecliptic is well displayed, and this means that it should be as high in the sky as possible. This avoids the difficulty of finding a clear horizon, and also overcomes the problem of atmospheric absorption, which greatly reduces the light of the stars. Thus a star at an altitude of 10 degrees suffers a loss of 60 per cent of its light, which corresponds to a whole magnitude; at an altitude of only 4 degrees, the loss may amount to two magnitudes.

The position of the ecliptic in the sky is therefore of great importance, and since it is tilted at about 23½ degrees to the Equator, it is only at certain times of the day or year that it is displayed to the best advantage. It will be realized that the Sun (and therefore the ecliptic) is at its highest in the sky at noon in midsummer, and at its lowest at noon in midwinter. Allowing for the daily motion of the sky, these times lead to the fact that the ecliptic is highest at midnight in winter, at sunset in the spring, at noon in summer and at sunrise in the autumn. Hence these are the best times to see the planets. Thus, if Venus is an evening object in the western sky after sunset, it will be seen to best advantage if this occurs in the spring, when the ecliptic is high in the sky and slopes down steeply to the horizon. This means that the planet is not only higher in the sky, but will remain for a much longer period above the horizon. For similar reasons, a morning object will be seen at its best on autumn mornings before sunrise, when the ecliptic is high in the east. The outer planets, which can come to opposition (i.e. opposite the Sun), are best seen when opposition occurs in the winter months, when the ecliptic is high in the sky at midnight.

The seasons are reversed in the Southern Hemisphere, spring beginning at the September Equinox, when the Sun crosses the Equator on its way south, summer beginning at the December

Solstice, when the Sun is highest in the southern sky, and so on. Thus, the times when the ecliptic is highest in the sky, and therefore best placed for observing the planets, may be summarized as follows:

	Midnight	*Sunrise*	*Noon*	*Sunset*
Northern lats.	December	September	June	March
Southern lats.	June	March	December	September

In addition to the daily rotation of the celestial sphere from east to west, the planets have a motion of their own among the stars. The apparent movement is generally *direct*, i.e. to the east, in the direction of increasing longitude, but for a certain period (which depends on the distance of the planet) this apparent motion is reversed. With the outer planets this *retrograde* motion occurs about the time of opposition. Owing to the different inclination of the orbits of these planets, the actual effect is to cause the apparent path to form a loop, or sometimes an S-shaped curve. The same effect is present in the motion of the inferior planets, Mercury and Venus, but it is not so obvious, since it always occurs at the time of inferior conjunction.

The inferior planets, Mercury and Venus, move in smaller orbits than that of the Earth, and so are always seen near the Sun. They are most obvious at the times of greatest angular distance from the Sun (greatest elongation), which may reach 28 degrees for Mercury, or 47 degrees for Venus. They are seen as evening objects in the western sky after sunset (at eastern elongations) or as morning objects in the eastern sky before sunrise (at western elongations). The succession of phenomena, conjunctions and elongations, always follows the same order, but the intervals between them are not equal. Thus, if either planet is moving round the far side of its orbit its motion will be to the east, in the same direction in which the Sun appears to be moving. It therefore takes much longer for the planet to overtake the Sun – that is, to come to superior conjunction – than it does when moving round to inferior conjunction, between Sun and Earth. The intervals given in the following table are average values; they remain fairly constant in the case of Venus, which travels in an almost circular orbit. In the case of Mercury, however, conditions vary widely because of the great eccentricity and inclination of the planet's orbit.

		Mercury	*Venus*
Inferior conj.	to Elongation West	22 days	72 days
Elongation West	to Superior conj.	36 days	220 days
Superior conj.	to Elongation East	36 days	220 days
Elongation East	to Inferior conj.	22 days	72 days

The greatest brilliancy of Venus always occurs about 36 days before or after inferior conjunction. This will be about a month *after* greatest eastern elongation (as an evening object), or a month *before* greatest western elongation (as a morning object). No such rule can be given for Mercury, because its distance from the Earth and the Sun can vary over a wide range.

Mercury is not likely to be seen unless a clear horizon is available. It is seldom seen as much as 10 degrees above the horizon in the twilight sky in northern latitudes, but this figure is often exceeded in the Southern Hemisphere. This favourable condition arises because the maximum elongation of 28 degrees can occur only when the planet is at aphelion (farthest from the Sun), and this point lies well south of the Equator. Northern observers must be content with smaller elongations, which may be as little as 18 degrees at perihelion. In general, it may be said that the most favourable times for seeing Mercury as an evening object will be in spring, some days before greatest eastern elongation; in autumn, it may be seen as a morning object some days after greatest western elongation.

Venus is the brightest of the planets and may be seen on occasions in broad daylight. Like Mercury, it is alternately a morning and an evening object, and it will be highest in the sky when it is a morning object in autumn, or an evening object in spring. The phenomena of Venus given in the table above can occur only in the months of January, April, June, August and November, and it will be realized that they do not all lead to favourable apparitions of the planet. In fact, Venus is to be seen at its best as an evening object in northern latitudes when eastern elongation occurs in June. The planet is then well north of the Sun in the preceding spring months, and is a brilliant object in the evening sky over a long period. In the Southern Hemisphere a November elongation is best. For similar reasons, Venus gives a prolonged display as a morning object in the months following western elongation in November (in northern latitudes) or in June (in the Southern Hemisphere).

The superior planets, which travel in orbits larger than that of the Earth, differ from Mercury and Venus in that they can be seen opposite the Sun in the sky. The superior planets are morning objects after conjunction with the Sun, rising earlier each day until they come to opposition. They will then be nearest to the Earth (and therefore at their brightest), and will then be on the meridian at midnight, due south in northern latitudes, but due north in the Southern Hemisphere. After opposition they are evening objects, setting earlier each evening until they set in the west with the Sun at the next conjunction. The change in brightness about the time of opposition is most noticeable in the case of Mars, whose distance from Earth can vary considerably and rapidly. The other superior planets are at such great distances that there is very little change in brightness from one opposition to another. The effect of altitude is, however, of some importance, for at a December opposition in northern latitudes the planets will be among the stars of Taurus or Gemini, and can then be at an altitude of more than 60 degrees in southern England. At a summer opposition, when the planet is in Sagittarius, it may only rise to about 15 degrees above the southern horizon, and so makes a less impressive appearance. In the Southern Hemisphere, the reverse conditions apply; a June opposition being the best, with the planet in Sagittarius at an altitude which can reach 80 degrees above the northern horizon for observers in South Africa.

Mars, whose orbit is appreciably eccentric, comes nearest to the Earth at an opposition at the end of August. It may then be brighter even than Jupiter, but rather low in the sky in Aquarius for northern observers, though very well placed for those in southern latitudes. These favourable oppositions occur every fifteen or seventeen years (1956, 1971, 1988, 2003) but in the Northern Hemisphere the planet is probably better seen at an opposition in the autumn or winter months, when it is higher in the sky. Oppositions of Mars occur at an average interval of 780 days, and during this time the planet makes a complete circuit of the sky.

Jupiter is always a bright planet, and comes to opposition a month later each year, having moved, roughly speaking, from one Zodiacal constellation to the next.

Saturn moves much more slowly than Jupiter, and may remain in the same constellation for several years. The brightness of Saturn depends on the aspects of its rings, as well as on the distance from Earth and Sun. The rings were inclined towards the Earth and Sun in 1980 and are currently near their maximum opening. The next passage of both Earth and Sun through the ring-plane will not occur until 1995.

Uranus, *Neptune*, and *Pluto* are hardly likely to attract the attention of observers without adequate instruments.

Phases of the Moon 1993

New Moon				*First Quarter*				*Full Moon*				*Last Quarter*			
	d	h	m		d	h	m		d	h	m		d	h	m
Jan.	22	18	27	Jan.	1	03	38	Jan.	8	12	37	Jan.	15	04	01
Feb.	21	13	05	Jan.	30	23	20	Feb.	6	23	55	Feb.	13	14	57
Mar.	23	07	14	Mar.	1	15	46	Mar.	8	09	46	Mar.	15	04	16
Apr.	21	23	49	Mar.	31	04	10	Apr.	6	18	43	Apr.	13	19	39
May	21	14	06	Apr.	29	12	40	May	6	03	34	May	13	12	20
June	20	01	52	May	28	18	21	June	4	13	02	June	12	05	36
July	19	11	24	June	26	22	43	July	3	23	45	July	11	22	49
Aug.	17	19	28	July	26	03	25	Aug.	2	12	10	Aug.	10	15	19
Sept.	16	03	10	Aug.	24	09	57	Sept.	1	02	33	Sept.	9	06	26
Oct.	15	11	36	Sept.	22	19	32	Sept.	30	18	54	Oct.	8	19	35
Nov.	13	21	34	Oct.	22	08	52	Oct.	30	12	38	Nov.	7	06	36
Dec.	13	09	27	Nov.	21	02	03	Nov.	29	06	31	Dec.	6	15	49
				Dec.	20	22	26	Dec.	28	23	05				

All times are G.M.T.
Reproduced, with permission, from data supplied by the Science and Engineering Research Council.

Longitudes of the Sun, Moon and Planets in 1993

DATE		*Sun*	*Moon*	*Venus*	*Mars*	*Jupiter*	*Saturn*
		°	°	°	°	°	°
January	6	286	73	332	108	194	317
	21	301	281	348	103	195	319
February	6	317	123	3	99	195	320
	21	332	326	14	99	194	322
March	6	345	131	19	101	193	324
	21	0	335	18	104	191	325
April	6	16	185	9	110	189	327
	21	31	20	4	117	187	328
May	6	45	223	7	124	186	329
	21	60	53	16	131	185	330
June	6	75	274	30	140	185	330
	21	90	101	44	148	185	330
July	6	104	308	60	157	187	330
	21	118	139	76	166	188	329
August	6	134	353	95	176	191	328
	21	148	193	112	185	193	327
September	6	163	37	131	196	196	326
	21	178	245	149	206	199	325
October	6	193	70	168	216	202	324
	21	208	281	186	226	206	324
November	6	224	118	206	238	209	324
	21	239	328	224	248	212	324
December	6	254	155	244	260	215	325
	21	269	0	262	270	218	326

Longitude of *Uranus* 290°
Neptune 290°

Moon: Longitude of ascending node
Jan. 1: 260° Dec. 31: 241°

Mercury moves so quickly among the stars that it is not possible to indicate its position on the star charts at a convenient interval. The monthly notes must be consulted for the best times at which the planet may be seen.

The positions of the other planets are given in the table on the previous page. This gives the apparent longitudes on dates which correspond to those of the star charts, and the position of the planet may at once be found near the ecliptic at the given longitude.

Examples

In the Southern Hemisphere two planets are seen low in the eastern morning sky in early November. Identify them.

> The southern star chart 2R shows the eastern sky at November 6^d05^h and shows longitudes 120°–210°. Reference to the table on page 71 gives the longitude of Venus as 206° and that of Jupiter as 209°, on November 6. Thus these planets are found low in the eastern sky and the brighter one is Venus.

The positions of the Sun and Moon can be plotted on the star maps in the same manner as for the planets. The average daily motion of the Sun is 1°, and of the Moon 13°. For the Moon an indication of its position relative to the ecliptic may be obtained from a consideration of its longitude relative to that of the ascending node. The latter changes only slowly during the year as will be seen from the values given on the previous page. Let us call the difference in longitude of Moon-node, d. Then if d = 0°, 180° or 360° the Moon is on the ecliptic. If d = 90° the Moon is 5° north of the ecliptic and if d = 270° the Moon is 5° south of the ecliptic.

On December 6 the Moon's longitude is given as 155° and the longitude of the node is found by interpolation to be about 242°. Thus d = 273° and the Moon is about 5° south of the ecliptic. Its position may be plotted on northern star charts 1L, 2L, 3L and southern star charts 1R, 2R.

Events in 1993

ECLIPSES

There will be four eclipses, two of the Sun and two of the Moon.

May 21: partial eclipse of the Sun – North America, Europe.
June 4: total eclipse of the Moon – western North America, Australasia, Asia.
November 13: partial eclipse of the Sun – South America, Australasia.
November 29: total eclipse of the Moon – Europe, Africa, America, Asia.

THE PLANETS

Mercury may be seen more easily from northern latitudes in the evenings about the time of greatest eastern elongation (February 21) and in the mornings around greatest western elongation (November 22). In the Southern Hemisphere the dates are April 5 (morning) and October 14 (evening).

Venus is visible in the evenings until March and in the mornings from April until December.

Mars is at opposition on January 7.

Jupiter is at opposition on March 30.

Saturn is at opposition on August 19.

Uranus is at opposition on July 12.

Neptune is at opposition on July 12.

Pluto is at opposition on May 14.

JANUARY

Full Moon: January 8 *New Moon:* January 22

EARTH is at perihelion (nearest to the Sun) on January 4 at a distance of 147 million kilometres.

MERCURY passes through superior conjunction on January 23 and is therefore too close to the Sun for observation throughout the month.

VENUS, magnitude −4.3, is a magnificent object in the south-western sky in the evenings, visible for several hours after sunset.

MARS reaches opposition on January 7, magnitude −1.4, and therefore is visible throughout the hours of darkness. Mars is easily located in the constellation of Gemini, because of its slightly reddish hue. It is moving slowly westwards, south of the Heavenly Twins, Castor and Pollux. The path of Mars amongst the stars during the first few months of the year is shown in Figure 1. At closest approach on January 3, it is 94 million kilometres from the Earth.

JUPITER, magnitude −2.1, is a brilliant morning object in Virgo and by the end of the month it is visible above the east-south-eastern horizon before midnight. The path of Jupiter amongst the stars is shown in Figure 6, given with the monthly notes for May.

SATURN begins the year as an evening object visible for a short while after dark low above the south-western horizon. During January it gradually gets more difficult to observe and around the middle of the month it is lost in the evening twilight.

ORBITS OF EARTH AND MARS. The Earth reaches perihelion – its closest point to the Sun – on January 4, at a distance of 147,000,000 kilometres or 91,400,000 miles. At aphelion, in early July, the

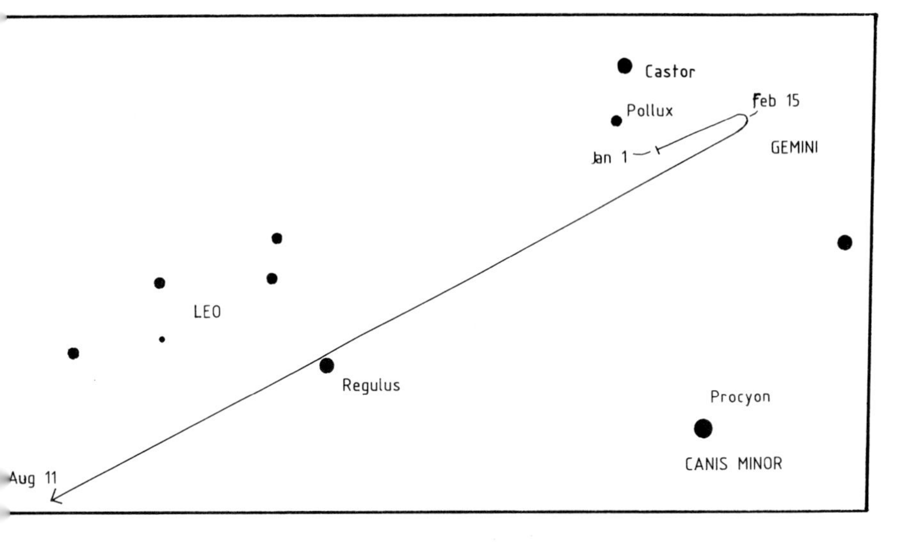

Figure 1. The path of Mars.

distance is 152,000,000 kilometres or 94,500,000 miles. This represents an orbital eccentricity of 0.017 – not so very different from a circle. The seasons have very little to do with this changing distance from the Sun. Theoretically the southern summers should be slightly shorter and hotter, with longer, colder winters, and this is true enough, but the effects are largely masked by the fact that there is so much more ocean in the southern hemisphere – and large expanses of water tend to stabilize temperatures.

The situation is different with Mars, which reaches opposition early this month – so that the Sun, the Earth and Mars are virtually lined up, with the Earth in the mid position. This year's opposition is not a favourable one, with the planet showing a maximum apparent diameter of less than 15 seconds of arc. The reason is that the Martian orbit is much less circular than ours. The orbital eccentricity is 0.093, and the perihelion and aphelion distances are, respectively, 206,700,000 km (128,500,000 miles) and 249,100,000 km (154,800,000 miles). This is a very marked difference. This year Mars comes to opposition almost at the time of its aphelion, so that the minimum distance between the two worlds is over 58,000,000 miles. At a perihelic opposition the maximum apparent diameter may exceed 25 seconds of arc – as will next happen in August 2003. (To British observers, there is some compensation in the fact that in 1993 Mars is in Gemini, about as far north in the sky as it can ever reach, while in 2003 it will be in Capricornus, well south of the celestial equator.)

The axial tilt of Mars is very similar to that of the Earth: 23°59′. This means that the seasons are basically of the same type, though they are much longer; the 'year' of Mars is equal to 687 Earth days, or 669 Mars days or 'sols' in view of the fact that Mars spins rather more slowly – the axial rotation period is 24 hours 37½ minutes. But there are no seas on Mars, so that the climate in the southern hemisphere is decidedly more extreme than that in the north. For example, the southern polar cap can become much larger than its northern counterpart can ever do. The Martian calendar is as follows:

Southern spring (northern autumn)	146 Earth days,	142 sols
Southern summer (northern winter)	160	156
Southern autumn (northern spring)	199	194
Southern winter (northern summer)	182	177
	687	669

Various Martian calendars have been drawn up. At the present these are academic exercises only, but if men reach Mars during the coming century, which appears to be extremely probable, an official calendar will certainly have to be internationally agreed!

CAPELLA. During January evenings the brilliant yellow star Capella (Alpha Aurigæ) is near the overhead point as seen from Britain and the northern United States. From England it is circumpolar, though when at its lowest – as in summer evenings – it almost grazes the horizon. The declination is (in round figures) +46 degrees. This means that from latitudes north of N.44° it never sets, and from latitudes south of S.44° it can never be seen at all. It can be seen from most inhabited countries, and this is a good time for South Africans and Australians to look for it, low over the northern horizon, but it is lost from the southern tip of New Zealand; from Invercargill (latitude −46°) it remains out of view. It is, in fact, one of three first-magnitude stars which cannot be seen from Invercargill; the other two are Vega and Deneb.

THIS MONTH'S ANNIVERSARY. Our 'anniversary astronomer' this month is not likely to be known to many people, though he carried out some useful work. Carl Johan Danielsson-Hill was Swedish, and was born in Målilla on January 12 1783. He was Lecturer in Physics at Lund University from 1819; Astronomer from 1821 to 1827, and Professor of Mathematics from 1830. He was primarily a mathematician, though he was also an active observer. He died on August 25 1875.

FEBRUARY

Full Moon: February 6 *New Moon:* February 21

MERCURY is an evening object after the first week or ten days of the month. For observers in northern temperate latitudes this will be the most favourable evening apparition of the year. Figure 2 shows, for observers in latitude N.52°, the changes in azimuth (true bearing from the north through east, south and west) and altitude of Mercury on successive evenings when the Sun is 6° below the horizon. This condition is known as the end of evening civil twilight, and in this latitude and at this time of year occurs about 35 minutes after sunset. The changes in the brightness of the planet are indicated by the relative sizes of the circles marking Mercury's position at five-day intervals. It will be noticed that Mercury is at its brightest before it reaches greatest eastern elongation (18°) on February 21. The magnitude of Mercury fades from −1.1 on February 12 to +1.0 by the end of the month.

VENUS continues to be visible as a magnificent evening object. It attains its greatest brilliancy (magnitude −4.5) on February 24 and completely dominates the western sky for several hours after sunset. Now is a good opportunity to look for Venus in daylight if you can shield your eyes from direct sunlight and know exactly where to look – Venus is 46° from the Sun at the beginning of February and 38° at the end.

MARS, magnitude −0.5, is still visible for the greater part of the night though now a whole magnitude fainter than when at opposition. Mars is in Gemini and already high in the eastern sky as soon as it is dark. As will be seen from the diagram given with the notes for January, Mars reaches a stationary point close to ε Geminorum, during the second half of February.

JUPITER continues to be visible as a morning object in Virgo, magnitudes −2.3.

SATURN remains too close to the Sun for observation as it passes through conjunction on February 9.

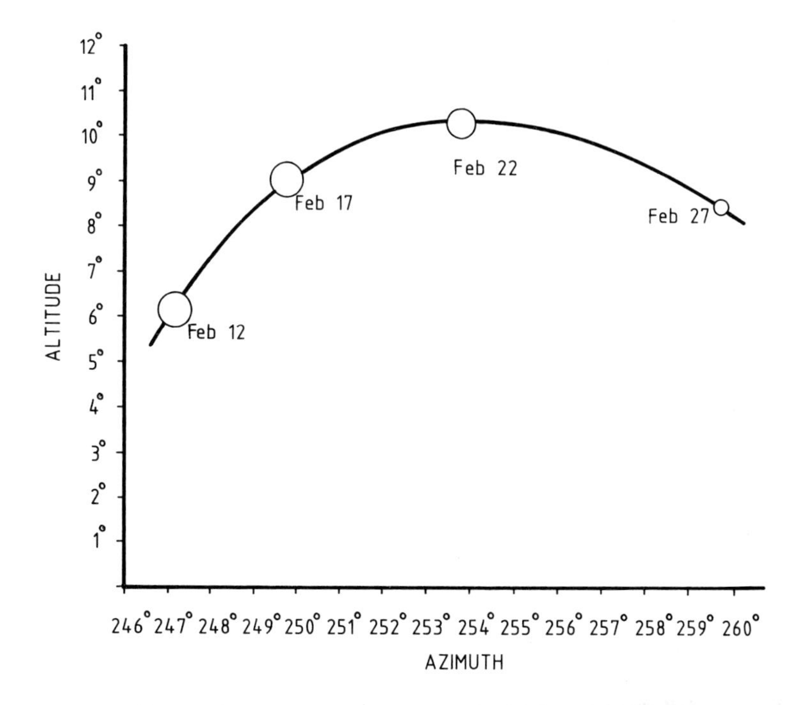

Figure 2. The evening apparition of Mercury for latitude N.52°.

PHASES OF MERCURY AND VENUS. Both the innermost planets are on view this month, and both reach their maximum brightness – but they behave differently. Mercury is at its most brilliant well before it reaches elongation, so appearing as a half-disk; Venus is at its best during its crescent stage.

It is obvious that as the phase of the planet shrinks, the apparent diameter increases. When 'full', Mercury and Venus are on the far side of the Sun, and are to all intents and purposes unobservable. They behave differently simply because Venus has a dense, cloud-laden atmosphere, while the Mercurian atmosphere is so rarefied that it can be ignored. The albedo (reflecting power) of Venus is therefore about 0.8, while that of Mercury is much less reflective – it sends back only 6 per cent of the sunlight falling upon it. This means that it is brightest when more of its surface is facing us.

With Venus, the time of theoretical dichotomy or half-phase never quite agrees with observation; during evening elongations dichotomy is early, during morning elongations it is late, sometimes by several days (this is known as Schröter's effect, since it was first pointed out by the pioneer planetary observer Johann Schröter almost two hundred years ago; the name was first used by the Editor of this *Yearbook*, but has now been generally accepted). The cause is again Venus's atmosphere. There should be no comparable Schröter effect for Mercury, but observations are much more difficult to make, and this month provides a good opportunity for observers who are adequately equipped.

MEBSUTA. This month Mars, now fading as it recedes from the Earth, passes fairly near the third-magnitude star Mebsuta or Epsilon Geminorum, in the Twins. Mebsuta lies in a position which means that it can even be occulted; indeed, this happened on April 7 1976, when Mars passed in front of the star. However, such phenomena are very rare indeed.

The name Mebsuta comes from the Arabic Al-Mabsutat, 'The Outstretched', because it marked the paw of the early Arabic Lion, but it is now on the hem of Castor's tunic! The name has been spelled in different ways: Meboula, Menita, Mesoula, and Mibwala are other variants, though in fact the name is hardly ever used. Proper names are in fashion for only the first-magnitude stars, plus a few other special cases such as Polaris, Mizar, and Mira.

Mebsuta has an apparent magnitude of 2.98, and a spectrum of type G8. The absolute magnitude – that is to say, the apparent magnitude that the star would have as seen from a standard distance of 32.6 light-years – is −4.5, so that Mebsuta would shine as brilliantly as Venus; but it is much further away than that, at a distance of 685 light-years. Look at it now, and you see it not as it is today, but as it was in the early part of the fourteenth century, when Edward I was King of England. The real luminosity is over 5000 times that of the Sun, so that Mebsuta is a giant star.

It has a faint companion of magnitude 9; the angular separation is 110 seconds of arc, and the position angle 094°. The secondary is not a difficult object, but since no relative motion has been found it is likely that the two components are not genuinely associated, so that

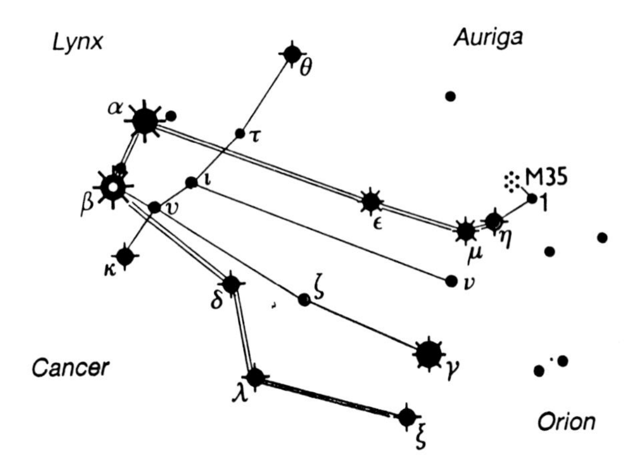

Figure 3. GEMINI – α = Castor; β = Pollux; ε = Mebsuta.

they do not make up a binary system – they merely happen to lie in almost the same line of sight as seen from Earth.

TWO CENTENARIES. Two eminent astronomers were born in 1893. One was M. G. J. Minnaert, whose birthday was February 12 of that year; he was born in Bruges, but spent most of his fruitful and (it must be said) often politically disturbed career at Utrecht University in Holland. He began his career as a biologist, graduating from Ghent in 1914, but went to Utrecht in 1921, and became a leading specialist in studies of the Sun. He was also active in the field of astronomical education, and a leading member of the International Astronomical Union. He died in 1970.

Megh Nad Saha was an outstanding Indian astrophysicist. His most important contribution was in the theory of ionization in stellar atmospheres, published in 1920. It was then generally chemical composition, but Saha showed that the depth of a spectral line depends instead upon the pressures and temperatures in the star's atmosphere. At the time of his death, on February 16 1956, Saha was a member of the Indian Parliament and Director of the Indian Association for the Cultivation of Science. He was, incidentally, very much concerned with calendar reform – which is understandable in view of the fact that there are many calendars in use in India!

MARCH

Full Moon: March 8 *New Moon:* March 23

Summer Time in Great Britain and Northern Ireland commences on March 28.

Equinox: March 20

MERCURY could possibly be detected as a difficult evening object, low on the west-south-western horizon after sunset, by keen sighted observers in the British Isles, on the first evening of the month. Mercury then passes through inferior conjunction on March 9. After the middle of the month observers in equatorial and southern latitudes should be able to observe it as a morning object and they should refer to the diagram (Figure 4) given with the notes for April. During this time its magnitude brightens from +2.0 to +0.6.

VENUS, magnitude −4.4, continues to be visible as a brilliant evening object in the western sky. However, observers will notice that the period available for observation shortens noticeably as Venus appears to move rapidly in towards the Sun, reaching inferior conjunction on April 1. Such a situation favours observers in the higher northern latitudes since Venus reaches its maximum northern ecliptic latitude towards the end of March. Thus observers in the British Isles will have the rare opportunity of seeing Venus as a morning object, low in the east-north-east before sunrise, several days *before* it passes through inferior conjunction. The optimum dates are the mornings of March 27–29. Although the planet is then only a mere 2 per cent illuminated the predicted magnitude is still as bright as −4.0.

MARS is still an evening object in Gemini, though its magnitude has faded to +0.2.

JUPITER, magnitude −2.4, is now visible throughout the hours of darkness, as it reaches opposition on March 30. At closest approach, one day after opposition, it is 666 million kilometres from the Earth.

SATURN remains too close to the Sun to be observed by those in high northern temperate latitudes. For those further south it gradually emerges from the morning twilight during the month to be visible for a very short while low in the south-eastern sky. Its magnitude is +0.8.

THE NAMES OF JUPITER'S SATELLITES. Jupiter is at opposition this month, and any small telescope (or even good binoculars) will show the four Galilean satellites. Until fairly recent times they were known only by their numbers, because apparently the now-familiar names of Io, Europa, Ganymede, and Callisto were suggested by Simon Marius, who claimed to have discovered the satellites before Galileo saw them (and may actually have done so). The fifth satellite was discovered by Barnard in 1892, and was the last planetary satellite to be found by visual observation; the name for it was suggested by the French astronomer Camille Flammarion, but again was not widely used for some time. It is a decidedly elusive object.

The remaining satellites, of course, are very faint indeed. The names for them, now endorsed by the International Astronomical Union and officially accepted, are all mythological. Metis was the goddess who prevented Saturn from swallowing unwelcome children; Adrastea was the nymph who suckled the infant Jupiter, and Amalthea the goat whose milk nourished the child; Lysithea was Jupiter's mother; Ananke was the personification of Fate; Carme, Thebe, Leda, Elara, and Himalia were among Jupiter's innumerable lovers (the morals of the Ancient Olympians were decidedly questionable); Sinope was a maiden whose rôle does not seem to be entirely clear, and Pasiphaë was the mother of the half-bull, half-man Minotaur who lived in the Cretan Labyrinth.

MARCH CENTENARIES. Two great astronomers have their centenaries this month. James Bradley was born in March 1693 at Sherborne in Dorset, and set out to study theology; he entered Balliol College, Oxford, in 1711 and graduated three years later. In 1719 he became a vicar in Bridstow, and was then appointed chaplain to the Bishop of Hertford, but by that time his astronomical interest had come to the fore, and he resigned his clerical position in 1721 to become Savilian Professor of Astronomy at Oxford. In 1742 he succeeded Edmond Halley as Astronomer Royal at Greenwich, retaining the position until his death, on July 13 1762, at Chalford in Gloucestershire.

Bradley's early observations were made at Wanstead Rectory with his uncle, the Rev. James Pound. They set out to measure stellar parallaxes, concentrating upon the star Gamma Draconis because it passes directly overhead and is thus particularly easy to measure. Most of the observations were made between 1725 and 1726 with S. Molyneux, at a private observatory in Kew. Parallax effects were not found – but the observations led Bradley on to the discovery of the aberration of light, which actually provided the first conclusive practical evidence of the Earth's motion round the Sun. Later, Bradley drew up a fundamental star catalogue, and also corrected the tables of Jupiter's Galilean satellites. Both as an observer and as a theorist he was outstanding.

Walter Baade was born on March 24 1893 at Shröttingshausen in Germany. He studied at the Universities of Münster and Göttingen, graduating in 1919; he then spent eleven years at the Bergedorf Observatory of Hamburg University. In 1931 he emigrated to America, to join the staff of the Mount Wilson Observatory. From 1948 to 1958 he worked at Palomar, but then returned to Germany to become Gauss Professor at Göttingen University. He died there on June 25 1960.

Baade was a skilled observer; in 1920 he discovered the strange asteroid Hidalgo, whose orbit takes it out as far as that of Saturn, and in 1948 he found Icarus, which is one of only two asteroids to approach the Sun within the orbit of Mercury (Phæthon is the other). On reaching Mount Wilson he worked with Edwin Hubble and Fritz Zwicky in research on supernovæ and the distances of galaxies. On the outbreak of war with Japan and Germany, Baade was technically an enemy alien, but was able to continue with his research – and even turn the wartime blackout to his advantage; in 1943 he was able to use the 100-inch reflector to study the Andromeda Galaxy, and this led him on to announce the discovery that there are two different 'stellar populations', Population I (mainly hot bluish stars in the spiral arms) and Population II (older, reddish stars near the centre of the system). This was a fundamental discovery, but later Baade followed it up with another of equal importance. Using the Palomar 200-inch reflector, he established that there are two different types of Cepheid variables with different period–luminosity laws.

Cepheid variables 'give away' their distances by the way in which they behave; the longer the period (that is to say, the interval between successive maxima) the more luminous the star. Hubble had found Cepheids, in the Andromeda Galaxy, and from their periods had deduced that the galaxy itself must be about 750,000 light-years away. Baade, however, found that Hubble's calculations had been worked out for Population II Cepheids – but in fact the stars concerned were of Population I, twice as luminous and therefore twice as remote. In one short paper, Baade doubled the size of the Universe. We now know that the distance of the Andromeda Galaxy is 2,200,000 light-years.

These were only two of Baade's many fundamental contributions to astronomy. Unquestionably he ranks with the leaders of twentieth-century research.

APRIL

Full Moon: April 6 *New Moon:* April 21

MERCURY is a morning object for observers in equatorial and southern latitudes. For observers in southern latitudes this will be the most favourable morning apparition of the year. Figure 4 shows, for observers in latitude S.35°, the changes in azimuth (true bearing from the north through east, south, and west) and altitude of Mercury on successive mornings when the Sun is 6° below the horizon. This condition is known as the beginning of morning civil twilight, and in this latitude and at this time of year occurs about 30 minutes before sunrise. The changes in the brightness of the planet are indicated by the relative sizes of the circles marking Mercury's position at five-day intervals. It will be noticed that Mercury is at its brightest after it reaches greatest western elongation (28°) on April 5. During the month its magnitude brightens from +0.6 to −0.6. Mercury is not visible to observers in the British Isles.

VENUS passes rapidly through inferior conjunction on April 1 and becomes visible as a brilliant morning object low in the eastern sky before dawn. Observers, except those in northern temperate latitudes, should note the proximity of Venus and Mercury around April 18 when Venus passes 7° north of Mercury and the thin crescent Moon is also near, a day later.

MARS, magnitude +0.8, is still visible as an evening object until around midnight (and well beyond for observers in northern temperate latitudes). Mars is now moving eastwards, passing south of Pollux in the middle of the month and then moving into Cancer.

JUPITER is only just past opposition and therefore still visible throughout the hours of darkness. Jupiter, magnitude −2.4, is in Virgo.

SATURN is a morning object, though for observers in the British Isles it is unlikely to be detected until fairly late in the month, low above the south-eastern horizon before twilight inhibits observation.

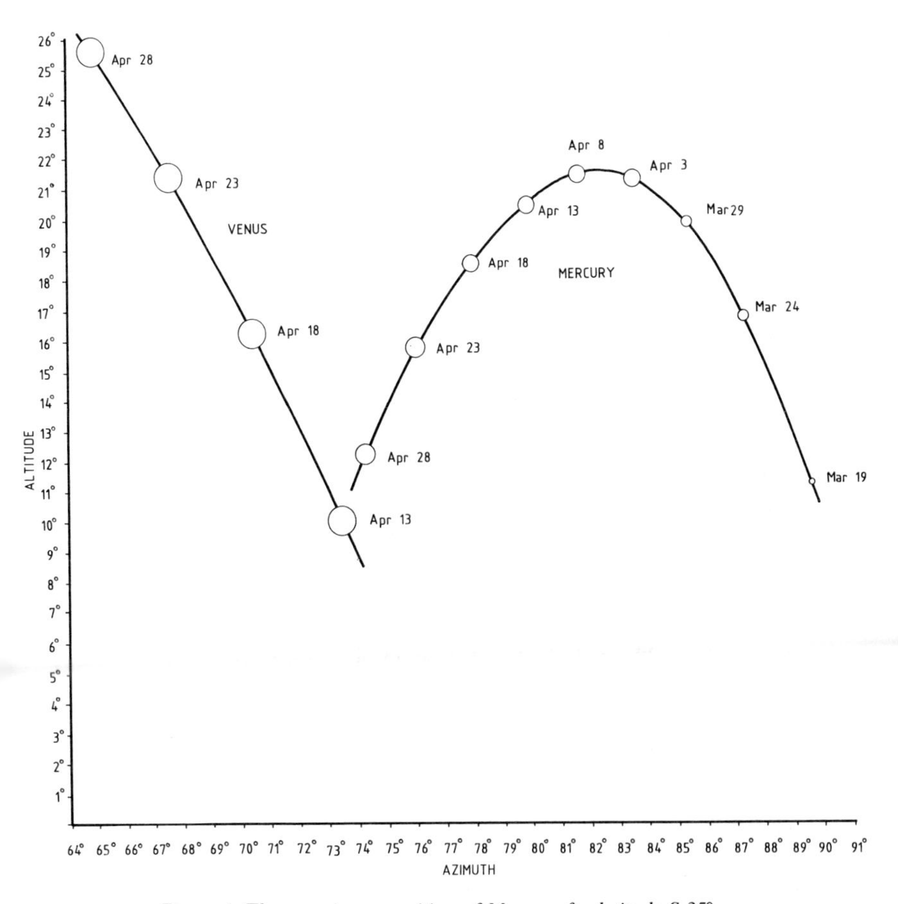

Figure 4. The morning apparition of Mercury for latitude S.35°.

THATCHER'S COMET. This month sees the return of the annual Lyrid meteor shower. The position of the radiant is R.A. 18^h 08^m, declination +32°. The shower begins around the 19th, reaches its maximum on the 21st, and ends on the 25th. Since the Moon is new on the 21st, 1993 is a good Lyrid year, and we may be in for a rich display. The average Zenithal Hourly Rate (ZHR) is around 10, but sometimes, as in 1922 and 1982, this is greatly increased. (The ZHR is the number of naked-eye meteors which would be expected to be seen with the naked eye by an observer under ideal conditions, with the radiant at the zenith. In practice these conditions are never fulfilled, so that the actual rate is always less than the ZHR; moreover, no account is taken of non-shower or sporadic meteors, which may appear from any direction at any moment.)

Most meteor showers, though not all, are associated with known comets, many of which are of short period. The Lyrids, however, are linked with a comet which comes back only once in several centuries. It was seen in 1861, and if the estimated period of 415 years is anywhere near the truth we may expect it again around AD 2276.

It was discovered on April 5 1861 by the American astronomer A. E. Thatcher, from New York. It was then described as a tailless nebulosity about 2 minutes of arc in diameter, with a central condensation; the magnitude was 7.5, below the limit of naked-eye visibility. As it drew inward it brightened, and by the end of the month was clearly visible without optical aid – indeed, an independent naked-eye discovery was made on April 28 by Backer in Germany. Between May 9 and 10 the comet passed less than 30,000,000 miles from the Earth; it was then of magnitude 2.5, with a degree-long tail. Perihelion was passed on June 3 at a distance of 86,000,000 miles from the Sun, slightly inside the Earth's orbit. It faded slowly, but was followed until September 7, when the magnitude had fallen to 10.

Obviously there is considerable uncertainty about the periods, but the orbit is certainly elliptical, so that Thatcher's Comet will return one day. Meanwhile, we can see its debris every year in the form of the April Lyrids.

URSA MAJOR. During April evenings Ursa Major, the Great Bear, is near the zenith in Britain and the northern United States; in America it is popularly called the Big Dipper, while in England it is more generally nicknamed the Plough. Its seven famous stars make up an unmistakable pattern. From countries such as South Africa and Australia it is always very low down, but it does rise, and April evenings are among the best times for looking for it. New Zealanders are less fortunate; from Wellington part of the Bear remains below the horizon, and almost the whole of it from the southern part of South Island.

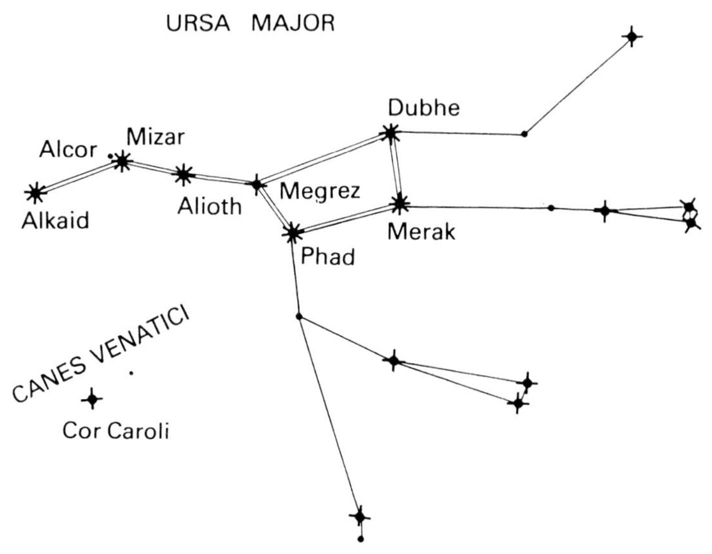

Figure 5. Ursa Major.

KARL HENCKE AND THE ASTEROIDS. Karl Ludwig Hencke was a typical last-century amateur astronomer. He was born on April 8 1793, and became postmaster at his native town, Driessen in Germany. He seems to have had no desire to become a professional scientist, but he will always be remembered for two discoveries he made – those of the asteroids Astræa and Hebe.

The first asteroid, or minor planet, was found by Piazzi on January 1 1801, the first day of the new century. He named it Ceres, in honour of the patron goddess of Sicily. Between that time and 1807 three more asteroids were found: Pallas, Juno, and Vesta. Ceres, with a diameter of 940 km or 584 miles, is much the largest of them, and in fact we now know it to be the giant of the entire swarm.

No more asteroids seemed to be forthcoming, and nothing more was done for some years. Then, in 1830, Karl Hencke decided to begin a systematic search. Patiently and alone he worked away, night after night, and at last he was rewarded. On December 8 1845 he came across a starlike point which proved to be Asteroid No. 5, Astræa. It was much smaller than the others, and has a diameter of no more than 120 km, or 75 miles, but it showed that there might be other bodies lurking in that part of the Solar System between the paths of Mars and Jupiter. Hencke kept searching, and on July 1 1847 he made his second and, as it turned out, final discovery: that of Asteroid No. 6, the 204-km or 127-mile Hebe.

Others joined in. Before the end of 1847 John Russell Hind had located Iris and Flora; Metis followed in 1848 and Hygeia in 1849. Since then no year has passed without the discovery of several more new asteroids. Hencke had no further success, but he had left his mark on the history of astronomy, and the modest postmaster of Driessen will not be forgotten. He died on September 21 1866.

MAY

Full Moon: May 6 *New Moon:* May 21

MERCURY, for observers in the latitudes of the British Isles, remains too close to the Sun for observation throughout the month, superior conjunction occurring on May 16. For observers further south the morning apparition of the planet, which commenced in March, continues through the first week of May, with a magnitude around −1. Towards the end of the month Mercury emerges from the evening twilight and is then visible low on the west-north-western horizon, again with a magnitude around −1.

VENUS, magnitude −4.4, is a magnificent object in the eastern sky before dawn, attaining its greatest brilliancy on May 7. For observers in the British Isles it is never visible for more than an hour before sunrise.

MARS continues to be visible as an evening object, magnitude +1.2, almost the same brightness as the two first-magnitude stars Pollux and Regulus as it passes between them around the middle of May. Figure 6 shows the path of Mars amongst the stars from August 11 to October 15.

JUPITER, magnitude −2.3, is a bright evening object in Virgo.

SATURN, magnitude +0.9, continues to be visible as a morning object, in Aquarius. Its path amongst the stars is shown in Figure 8, given with the monthly notes for June.

JOHN HERSCHEL AND GAMMA VIRGINIS. One of the most interesting binary stars in the sky is Gamma Virginis, which has several proper names: Arich, Porrima, and Postvarta. Since it lies less than two degrees south of the celestial equator it can be seen from every inhabited country, and it is well on view during evenings in May. It

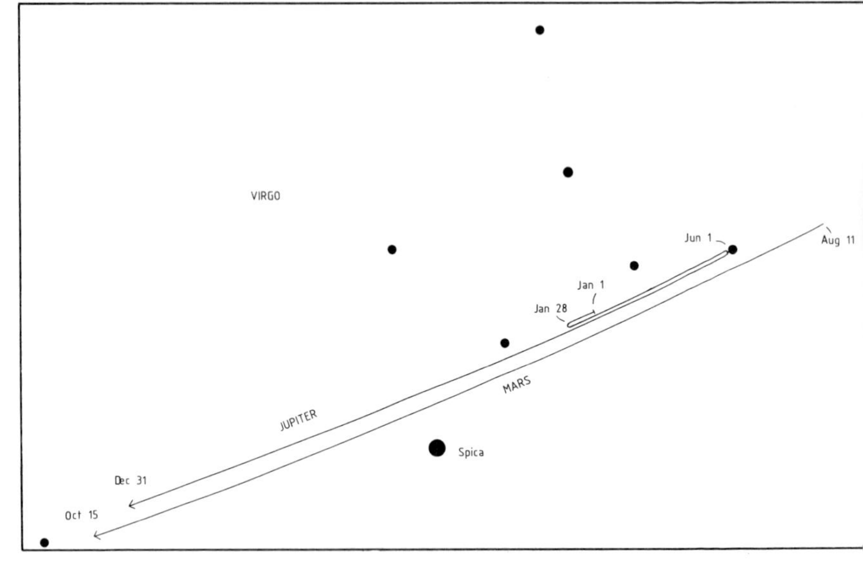

Figure 6. The paths of Mars and Jupiter.

is one of our nearer neighbours, at a distance of 32 light-years. There are two virtually identical components, each of spectral type F0 and each about 7 times as luminous as the Sun. Many observers call them yellow; others will see no colour in them.

The revolution period is 171.4 years, and the orbit is decidedly eccentric, so that as seen from Earth the angular separation changes considerably; at its greatest it is 6.2 seconds of arc, as in 1920, while at its least the components are so close together that in any ordinary telescope they appear as one star.

Gamma Virginis was the second binary system to have its orbit worked out (the first was Xi Ursæ Majoris, whose orbit was computed in 1828 by Savary). With Gamma Virginis, the calculations were undertaken by Sir John Herschel. In 1832 he published the following statement with regard to his measurements:

> 'If they be correct, the latter end of the year 1833, or the beginning of the year 1834, will witness one of the most striking phenomena which sidereal astronomy has yet afforded, viz. the perihelion passage of one star round another, with the immense

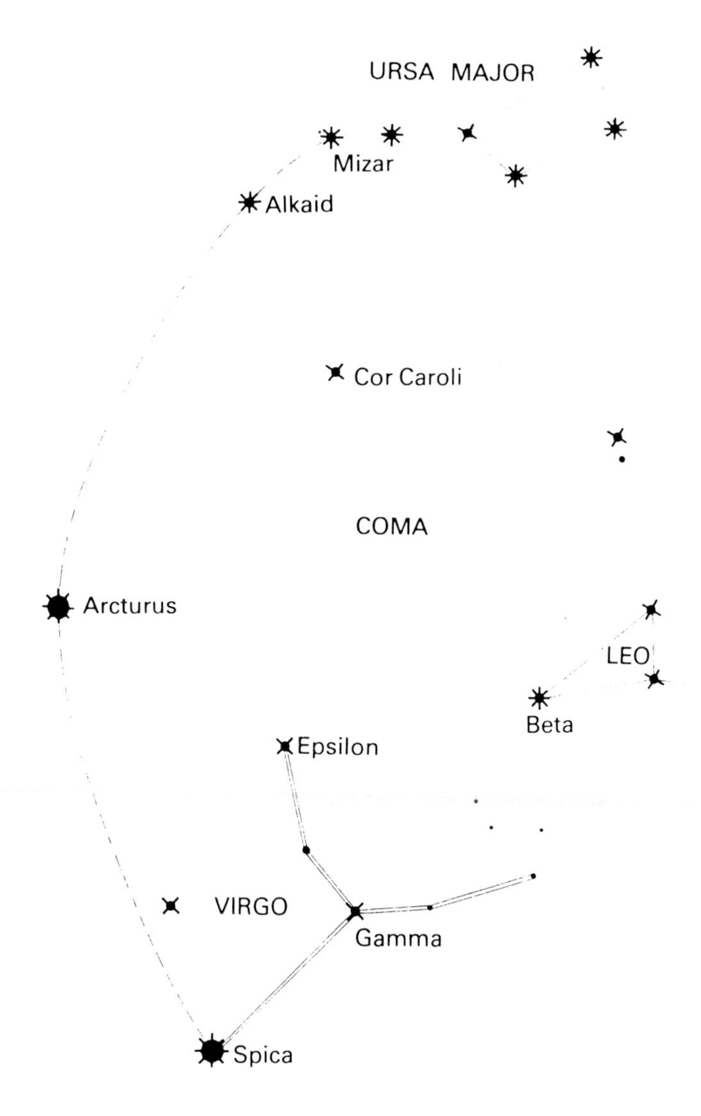

Figure 7. Gamma Virginis.

angular velocity of between 60 and 70 degrees per annum, that is to say, of a degree in 5 days. As the two stars will then, however, be within little more than half a second of each other, and as they are both large, and nearly equal, none but the very finest telescopes will have any chance of showing this magnificent phenomenon.'

The closest approach actually occurred in 1836. Three observers were following it: Admiral W. H. Smyth at Bedford in England, F. G. W. Struve at Dorpat in Estonia, and Sir John Herschel himself, who had then taken his 20-foot reflector to the Cape of Good Hope and was busy undertaking the first really detailed survey of the far-southern stars which can never be seen from Europe. Smyth wrote in January 1836 that 'instead of the appulse which a careful projection had led me to expect, I was astonished to find it a single star! In fact, whether the real disks were over each other or not, my whole powers, patiently worked from 240 to 1200, could only make the object round.' At Dorpat, Struve was able to use the best telescope in the world at that time – the Fraunhofer refractor – and using a magnification of ×848 could see that Gamma Virginis was elongated, though he could not split it. From the Cape, Herschel recorded on February 27 1836 that:

'Gamma Virginis, at this time, is to all appearance a single star. I have tormented it, under favourable circumstances, with the highest powers I can apply to my telescopes, consistently with seeing a well-defined disk, till my patience has been exhausted; and that lately, on several occasions, whenever the definition of stars generally, in that quarter of the heavens, would allow of observing with any chance of success, but I have not been able to procure any decisive symptom of its consisting of two individuals.'

Within a few months the two components could be separated again. Greatest separation occurred in 1920. Since then the pair has been closing, and though it is still easy enough the star will again appear single, except in giant telescopes, in the year 2007. This does not, of course, indicate that the components have really closed up; everything depends upon the angle from which we view them. The real separation between the two ranges between about 279,000,000 miles at periastron out to about 6,500,000,000 miles at apastron.

Both components have been suspected of being slightly variable; thus Struve wrote that between 1825 and 1831 the star now officially classed as the secondary appeared the brighter, but that in 1851 the converse was true. The reality of these changes is very doubtful, but observers may care to check whether there really are any relative changes. All in all, the two appear as perfect twins.

THIS MONTH'S ECLIPSE OF THE SUN. The track of the May 21 eclipse crosses northern Britain, so that would-be observers from the south will have to trek northward; the eclipse will be seen from most of the United States. Less than three-quarters of the solar disk will be covered by the Moon, and of course there is no chance of seeing the corona or prominences. However, eclipses of any kind are interesting to watch. As always, great care should be taken.

It is unsafe to stare at the Sun even with the naked eye, and on no account look at it through any telescope or binoculars. Dark filters are no sure protection.

The only sensible method to view the eclipse with a telescope is to project the Sun's image on to a screen held or fastened behind the eyepiece.

JUNE

Full Moon: June 4 *New Moon:* June 20

Solstice: June 21

MERCURY is an evening object, its magnitude fading during the month from −0.6 to +1.6. It should be looked for at the end of evening civil twilight, above the west-north-western horizon. Although Mercury attains its greatest eastern elongation on June 17 it is a difficult object for observers in the British Isles because of the long duration of twilight around the summer solstice. Their only hope of seeing the planet would be during the first half of the month.

VENUS, magnitude −4.2, continues to be visible as a brilliant morning object in the eastern sky before dawn. It reaches greatest western elongation (46°) on June 10.

MARS, magnitude +1.4, is visible in the western sky in the evenings, in Leo, passing north of Regulus on June 22.

JUPITER continues to be visible as an evening object in Virgo, magnitude −2.1. By the end of June it is lost to view over the south-western horizon before midnight.

SATURN continues to be visible as a morning object, magnitude +0.8. Observers in the British Isles will find it getting easier for them to see: for them, it is becoming visible above the south-eastern horizon before midnight, by the end of June. Figure 8 shows the path of Saturn amongst the stars throughout the year.

CHAMBERS' RED STARS. In 1895 G. F. Chambers published the third edition of his *Handbook of Astronomy*, which was (and still is) regarded as something of a classic. In it he included a catalogue of

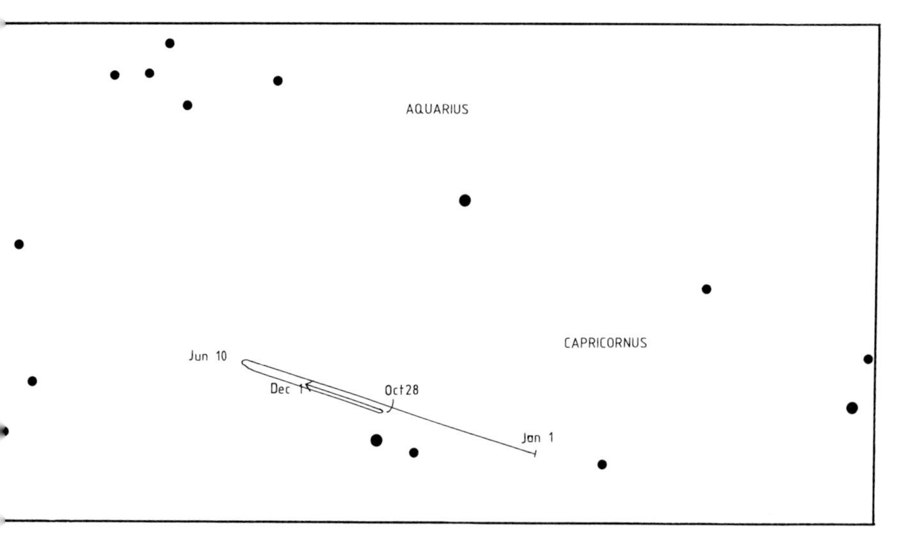

Figure 8. The path of Saturn.

red stars, and some of the entries are rather surprising. Of course, spectroscopy in those days was still in its relative infancy, and many estimates depended on the naked eye alone; all the same, it is interesting to look back at the various entries. In this list I have included only the stars above magnitude 5.3 which were classed as something more pronounced than orange. The magnitudes and spectral types are, of course, taken from modern catalogues.

Description (Chambers)	*Star*	*Mag.*	*Spectrum*
Intense reddish orange	Mu Cephei	3–6*v*	M2
Deep reddish orange	Alpha Tauri	0.8	K5
Fiery orange	Delta² Lyræ	4.3	M4
	Gamma Draconis	2.2	K5
	Pi Herculis	3.2	K3
Bright orange red	u Carinæ	3.8	K0
Fine deep orange	Zeta Aurigæ	3.7*v*	K4
Orange red	Delta Sagittarii	2.7	K2
Reddish yellow	Beta Doradûs	3.7*v*	F9
	Psi Pegasi	4.7	M3

Reddish orange	Pi Aurigæ	4.3	M3
	Theta Canis Majoris	4.1	K4
	Beta Cancri	3.5	K4
	Mu Geminorum	2.9	M3
	Mu Hydræ	3.8	K4
	Alpha Lyncis	3.1	M0
	Lambda Leonis	4.3	K5
	Kappa Libræ	4.7	K5
	Delta Ophiuchi	2.7	M1
	Alpha Orionis	0.5*v*	M2
	Omicron[1] Orionis	4.7	M3
Orange red	Alpha Crateris	4.1	K0
Reddish	Beta Gruis	2.1	M3
	Psi Ursæ Majoris	3.0	K1
Pale red	119 Tauri	4.3	M2
Decided red	Theta Ursæ Minoris	5.3	K4
Red and variable	L^2 Puppis	2.6–6.2	M
Very red	Mu Velorum	2.7	G5
	Phi[1] Lupi	3.6	K5
Fiery red	Mu Canis Majoris	5.2	M0
	Alpha Scorpii	1.0	M1

Finally, Eta Canis Majoris (2.4, B5) is given as 'perhaps purplish'. A notable omission is Gamma Crucis (1.6, M3) which is, however, so far south that it may have been insufficiently observed at that time. Also excluded is Mira Ceti, which is of course below naked-eye visibility for much of each year. It is surprising that Mu Cephei, the celebrated 'Garnet Star' – usually regarded as much the reddest of all the naked-eye stars – is listed as being less extreme than Antares or Mu Canis Majoris!

At about the same time two reports were published by the short-lived Star Colour Section of the British Astronomical Association. The only stars which were given as definitely red were Mira, Mu Cephei, and Antares; Betelgeux was given as orange, and Alphard or Alpha Hydræ as orange-yellow. Eye estimates are notoriously subject to personal acuity and colour vision, but all the same some of the discrepancies are curious. Few people will see much colour in the Cepheid variable Beta Doradûs, for example.

Antares, the red supergiant, is well placed this month during evenings – very high from southern countries, where Scorpius dominates the scene, but depressingly low from the British Isles;

part of the Scorpion never rises at all. The 'sting', marked by the two bright stars Lambda and Upsilon Scorpii (Shaula and Lesath), is very difficult to see from any part of England, and impossible from Scotland. This is a pity, because the Scorpion is such a magnificent constellation, and is also one of the few to give some sort of impression of the object after which it is named. Preceding it is the much less imposing Zodiacal constellation of Libra, the Scales or Balance, which was formerly known as Chelæ Scorpionis (the Scorpion's Claws); incidentally, the star once known as Gamma Scorpii has been given a free transfer by edict of the International Astronomical Union, and is now officially known as Sigma Libræ.

THIS YEAR'S LUNAR ECLIPSES. There are two total eclipses of the Moon this year: one on June 4 and the other on November 29. They are visible from different parts of the world, but some observers may have the chance to see them both, and it will be interesting to compare them. The brightness (or darkness!) of the eclipsed part of the lunar surface is very variable, according to the condition of the Earth's atmosphere at the time, and it is on record that on a few occasions the totally eclipsed Moon has vanished so completely that it could not be found even with a telescope – whereas during the 1848 eclipse the eclipsed disk was so bright that uninformed observers refused to believe that an eclipse was taking place at all.

JULY

Full Moon: July 3 *New Moon:* July 19

EARTH is at aphelion (furthest from the Sun) on July 4 at a distance of 152 million kilometres.

MERCURY passes rapidly through inferior conjunction on July 15. As a consequence, observers in equatorial and southern latitudes will continue to be able to see it as an evening object for the first week of the month. For about the last ten days of July they will be able to see it as a morning object, low above the eastern horizon before twilight inhibits observation. During this time its magnitude brightens from +3.0 to +0.9. For observers in the British Isles the long duration of twilight renders Mercury unobservable.

VENUS is still a brilliant morning object, magnitude −4.0, visible in the eastern sky before dawn.

MARS is still an evening object in Leo, magnitude +1.6. For observers in the latitudes of the British Isles it is coming towards the end of its period of visibility, being observable for only a short while low in the west-south-western sky before being lost in the long twilight.

JUPITER, magnitude −1.9, remains an evening object in Virgo, visible in the south-western sky.

SATURN is a morning object in Aquarius, magnitude +0.6, but now visible for the greater part of night.

URANUS is at opposition on July 12, in Sagittarius. The planet is only just visible to the naked eye under the best of conditions since its magnitude is +5.6. In a small telescope it appears as a slightly greenish disk. At opposition Uranus is 2781 million kilometres from the Earth. Uranus overtakes Neptune this year and there are three conjunctions of the two planets. If you define conjunction as having the same longitude then the dates are February 2, August 20, and October 24 but if you define it as having the same right ascension then the dates are considerably different, namely January 26, September 17, and September 28. You pay your money and you take your choice! On each occasion Uranus passes one degree south of Neptune.

NEPTUNE is at opposition on the same day as Uranus and, of course, is also in Sagittarius. It is not visible to the naked eye since its magnitude is +7.9. The distance of Neptune from the Earth at opposition is 4363 million kilometres.

THE BLUE GIANT. In August 1989 the American space-craft Voyager 2 flew past Neptune. It had been launched in 1977, and had already surveyed Jupiter, Saturn and Uranus; during the Neptune encounter it performed perfectly, and may well lay claim to being the most successful of all interplanetary probes.

The Voyager mission gave us our first really reliable information about Neptune, which proved to be a much more dynamic world than its rather bland twin, Uranus. Of the two, Neptune is very slightly the smaller, but is appreciably more massive. It has a gaseous surface, and is of a glorious blue colour. There may well be a silicate–iron core, but the interior of the globe is dominated by planetary 'ices', mainly water, and even the core may not be strongly differentiated from the ice components.

Radio measures fixed the rotation period of the core as 16 hours 3 minutes, rather shorter than had been anticipated. The rotation of the surface features is somewhat complicated, with a greater difference between the various latitudes than is the case with the other giants; moreover, Neptune is the 'windiest' planet in the Solar System. The main surface feature is the Great Dark Spot, which has a rotation period of 18.3 hours, and has an east–west diameter averaging about 10,000 miles or 16,000 kilometres – much greater than the diameter of the Earth. It seems to be a giant eddy, and is probably long-lived; high above it are wispy clouds of 'methane cirrus'. Further south is a smaller, very variable dark spot, which has a quicker rotation period and has been nicknamed the Scooter; still further south is another dark spot, which 'laps' the Great Dark Spot once in every five days.

Temperature measurements from Voyager showed that there is a cold mid-latitude region, with the pole and equator rather warmer. Of course, all the temperatures are very low – around minus 220 degrees Centigrade – but Neptune is no colder than Uranus, because, unlike its twin, it has considerable internal heat.

Before the Voyager mission Neptune was known to have two satellites, Triton (large) and Nereid (small); partial rings had been suspected. Voyager showed that there is a true ring system, and also discovered six new inner satellites, now named Naiad, Thalassa, Despina, Galatea, Larissa and Proteus. Proteus, the largest of them, has a diameter greater than that of the eccentric-orbit Nereid, but is not detectable from Earth because it is so much closer to Neptune.

Triton, the only large satellite, was already known to be exceptional inasmuch as it moves round Neptune in a retrograde direction – opposite to that in which Neptune itself spins. Instead of being rather larger than our Moon, as had been expected, it turned out to be rather smaller, with a diameter of 1681 miles (2705 kilometres). Moreover, the atmosphere, which had been thought to be reasonably dense, proved to be very tenuous indeed – 100,000 times thinner than our own air – and to be made up largely of nitrogen. The surface is very varied, but there is a general coating of ice, presumably water (H_2O) overlaid by nitrogen and methane ices; water ice has not been detected spectroscopically, but must exist, because nitrogen and methane ices are not hard enough to maintain surface relief over long periods. Normal craters are scarce.

The southern polar region, now known as Uhlanga Regio, is covered with pink nitrogen snow. In it have been found nitrogen ice geysers, which were completely unexpected. Their origin is not definitely known, but several plausible theories have been proposed. It seems that below the surface there is a layer of liquid nitrogen; if any of this migrates toward the surface through a weak point, the pressure is relaxed, and the result is an outrush of nitrogen ice and vapour. Material taken with the outrush is wafted downwind in the thin Tritonian atmosphere, producing dark streaks which are clearly visible in the Voyager pictures. It may be that the geysers are driven by heat from Triton's interior. Alternatively, there may be regions where there are areas of 'clear' nitrogen ice, so that the weak sunlight can penetrate through and build up the temperature, thereby acting rather in the manner of 'solar collectors' and storing energy in the form of gas pressure. Eventually the pressure causes a rupture in the outer layer, and a geyser results.

What we would like to do is to re-survey Neptune and Triton in the near future, and see what changes have taken place. Unfortunately this is out of the question. Even the Hubble Space Telescope (which is working excellently in many respects, despite its faulty mirror) cannot provide enough resolution for this. We must await the arrival of a new space-craft, and at the moment no plans have been made; moreover, NASA's funds have been savagely cut back, and after the collapse of the Soviet Union it is not easy to forecast the future of Russian space research.

There is another problem, too. Voyager 2 went to Neptune by using the 'gravity assist' technique, using the pulls of Jupiter, Saturn, and Uranus. In the late 1970s the four giant planets were suitably arranged in a curve, making it possible to send a probe from the one to the other; this will not recur for well over a century to come, so that a mission to Neptune will take longer than with Voyager 2. When we will next obtain close-range information remains to be seen. For the moment, we can only be patient.

AUGUST

Full Moon: August 2 *New Moon:* August 17

MERCURY continues its morning apparition for observers in equatorial and southern latitudes, though by the middle of the month it is getting too close to the Sun for observation as Mercury moves towards superior conjunction on August 29.

VENUS, magnitude −3.9, continues to be visible as a brilliant morning object, visible in the eastern sky for several hours before dawn.

MARS, magnitude +1.7, is now at its faintest magnitude for 1993. It is in Virgo and still visible as an evening object, though observers in northern temperate latitudes will be losing it in the twilight – in fact observers as far north as the British Isles will lose it at the very beginning of August and will not see it again this year.

JUPITER continues to be visible as a bright evening object, magnitude −1.7, in the south-western sky after sunset. For observers in the British Isles it is only visible for a short while before setting and by the end of August it will be a difficult object to detect.

SATURN, magnitude +0.4, is visible throughout the hours of darkness as it comes to opposition on August 19 at a distance of 1316 million kilometres. Saturn is in Aquarius.

THE GLOBULAR CLUSTERS IN HERCULES. In mythology, Hercules was a great hero, but in the sky he is, frankly, rather obscure. The constellation covers a wide area, and is well on view this month; though well north of the celestial equator, it still attains a respectable altitude above the South African or Australian horizon. There are only two stars above the third magnitude, Beta (Kornephoros) and Zeta (Rutilicus), each of magnitude 2.8, though the red supergiant variable Alpha (Rasalgethi) can reach 3.0 when at its maximum. Hercules occupies the large triangle bounded by Vega, Rasalhague (Alpha Ophiuchi) and Alphekka (Alpha Coronæ Borealis).

The most notable objects are the two globular clusters, M.13 and M.92. M.13 is the brightest globular accessible from Britain; it is surpassed only by the two southern systems of Omega Centauri and 47 Tucanæ. It was discovered in 1714 by Edmond Halley; on the whole it is rather surprising that it was not found earlier, because it is just visible with the naked eye as a misty speck between Zeta and

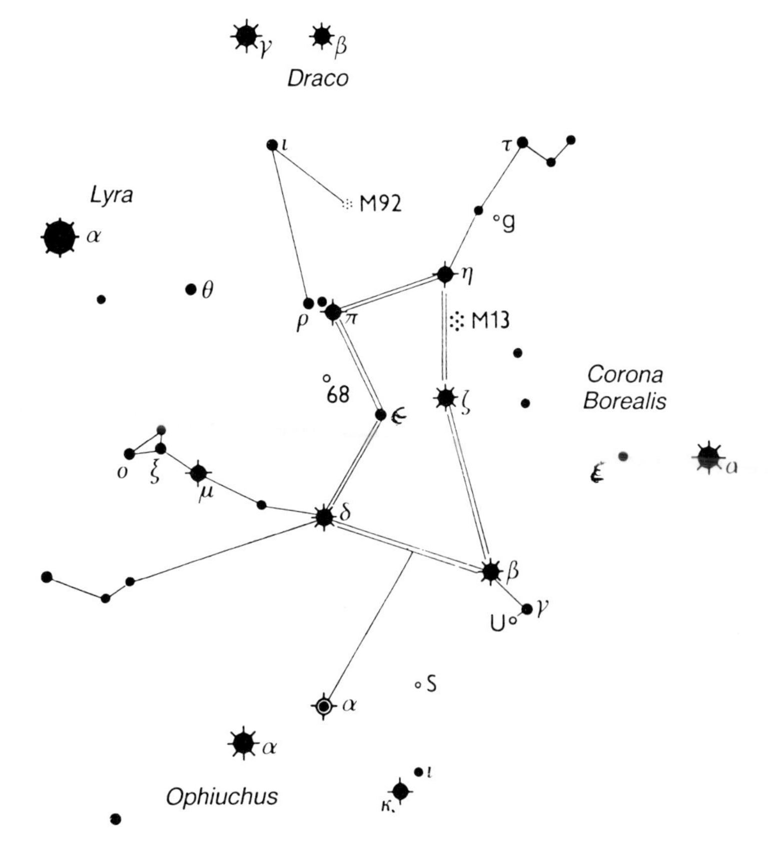

Figure 9. The globular clusters in Hercules.

Eta Herculis. Halley recorded that 'This is but a little patch, but it shows itself to the naked eye when the sky is serene and the Moon absent', while Messier, in 1764, commented that it was 'a nebula which I am sure contains no star; round and brilliant, centre brighter than the edges'. In fact, the outer parts are not hard to resolve with a small modern telescope. Its distance is approximately 22,500 light-years. Oddly enough, it is rather poor in short-period variable stars.

The second of the Hercules globulars, M.92, was discovered by J. E. Bode in 1777. It is smaller than M.13, with a diameter of about 90 light-years as against 160 light-years, and it is also further away; the distance is around 35,000 light-years. It is therefore fainter, but it is still on the fringe of naked-eye visibility, and some people with very keen sight can detect it. It is, of course, easy in binoculars, and with a magnification of ×8 it is in the same field with Iota Herculis and the Pi-Rho pair. All in all it is not greatly inferior to M.13, though it is smaller and appreciably more condensed. Messier, in 1781, wrote that it was 'a fine, distinct and very bright nebula . . . the centre is clear and bright, surrounded by nebulosity, and it resembled the nucleus of a large comet'. Like M.13 it is relatively poor in short-period variable stars, though numbers of them have now been found.

THE AUGUST METEORS. The Perseid meteor shower reaches its maximum on August 12, and fortunately moonlight does not greatly interfere this year. Much interest has been centred on the parent comet, Swift–Tuttle, which has been seen at only one return: that of 1862, when it reached the second magnitude and developed a 30-degree tail. The period was given as 120 years. In this case the comet should have returned in 1982, but despite intensive searches it failed to put in an appearance. Since then there have been suggestions that it may be identical with Kegler's Comet of 1737, which moved in a rather similar way and also became a prominent naked-eye object. If so, then the period may be more like 130 years, with a possible return in 1992. At the time when these words are being written (June 1992) the comet has not been located, but observers continue to hope!

SEPTEMBER

Full Moon: September 1 and 30 *New Moon:* September 16

Equinox: September 23

MERCURY, after the first ten days of the month, becomes visible as an evening object, low in the western sky around the time of evening civil twilight, though only for observers in equatorial and southern latitudes. Observers in these latitudes will notice a slight change in the brightness of Mercury as its magnitude changes from −0.7 to −0.1 during the period.

VENUS is a brilliant morning object, magnitude −3.8. It is well placed for observation in the eastern sky for an hour or so before sunrise.

MARS, magnitude +1.6, is still an evening object except for observers in northern temperate latitudes. Even for other observers it is only visible for a short while low in the western sky, after darkness falls.

JUPITER has already disappeared from view for observers in the latitudes of the British Isles. Observers further south may be able to detect it low in the south-western sky after sunset for the first part of the month but it gradually gets more and more difficult to observe. Mars and Jupiter are within 1° of each other around September 6–7, Jupiter being about three magnitudes brighter than Mars.

SATURN, not long past opposition, is visible for the greater part of the night. Its magnitude is +0.4.

FOMALHAUT. So far as British observers are concerned, September evenings provide the best time for locating Fomalhaut in Piscis

Austrinis (the Southern Fish), which with its declination of approximately −30 degrees is the southernmost of the first-magnitude stars to be visible from the British Isles. Piscis Austrinis – alternatively known as Piscis Austrinus (as on the diagram) – is one of Ptolemy's original forty-eight constellations, but apart from Fomalhaut it has no star above the fourth magnitude.

Fomalhaut's name comes from the Arabic Fum al Hūt, the Fish's Mouth. According to the French astronomer Camille Flammarion, the Persians of around BC 3000 called it *Hastourang* one, of the Guardians of Heaven and a Royal Star; it was then near the winter solstice. About BC 500 it is said to have been the object of sunrise worship at the Temple of Demeter at Eleusis; the Chinese knew it as Po Lo Sze Mun.

The best way to identify Fomalhaut is to use two of the stars in the Square of Pegasus (Beta or Scheat, and Alpha or Markab) as pointers; they show the way down to Fomalhaut, not far above the horizon. (Do not confuse Fomalhaut with Diphda or Beta Ceti, which is a magnitude fainter and, as seen from Britain, higher up.) Fomalhaut is easy to find from most of England, but less so from Scotland, and anyone living in, say, Aberdeen will be lucky to glimpse it at all. From southern countries it is, of course, high up, and those who see it at a reasonable altitude for the first time are often surprised at its brightness. In fact the magnitude is 1.2, and in the list of bright stars it comes eighteenth – much the same as Pollux in the Twins, and slightly brighter than Deneb in Cygnus, the Swan.

Fomalhaut is the nearest of all the first-magnitude stars apart from Sirius, Procyon, Altair, and Alpha Centauri. Its distance is 22 light-years; the spectral type is A3, so that it is pure white, and the luminosity is 13 times that of the Sun, roughly half that of Sirius.

In 1983 the Infra-Red Astronomical Satellite (IRAS) carried out a survey of the sky at infrared wavelengths, and some very interesting discoveries were made. In particular, it was found that some stars are associated with marked 'infrared excesses', indicating the presence of cool, possibly planet-forming material. Vega was the first of these; Fomalhaut is another, though the most impressive case of all is that of the fainter star Beta Pictoris, where the material has actually been photographed.

Does Fomalhaut, then, have a planetary system? There seems no definite reason why it should not; it is a stable, single star, and though it is much more luminous than the Sun it still ranks as a dwarf. It is very dangerous to jump to conclusions, but it is fair to

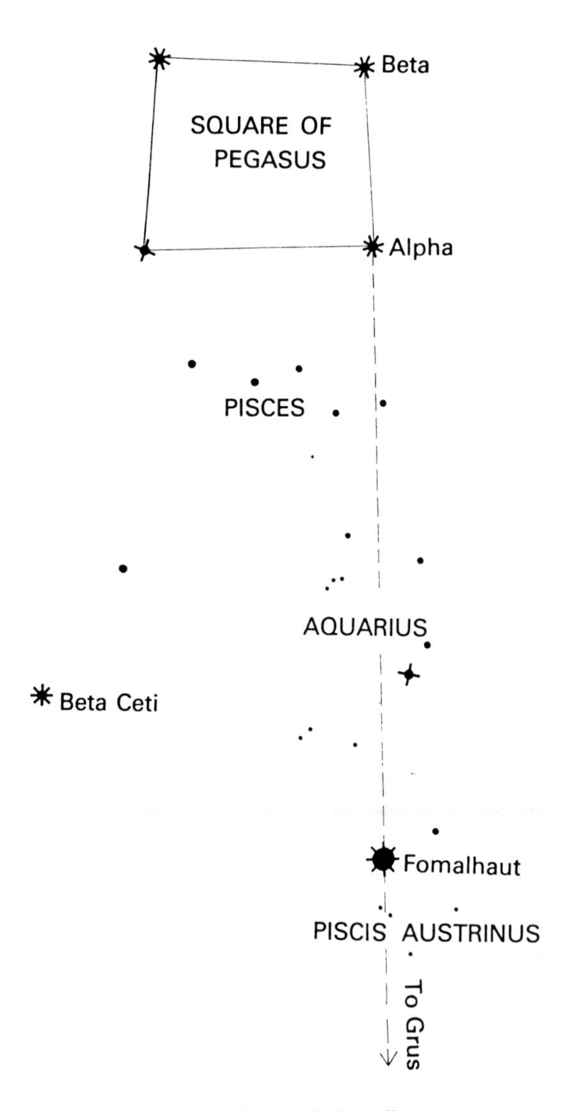

Figure 10. Fomalhaut.

say that the presence of planets round Fomalhaut cannot be ruled out.

If the line from Pegasus through Fomalhaut is prolonged, it will reach Grus, the Crane, which is much the most prominent of the four 'Southern Birds' (the others are Tucana, the Toucan; Pavo, the Peacock; and Phœnix, the Phœnix). Its brightest star, Alnair, has a declination of −47°.

OCTOBER

Full Moon: October 30 *New Moon:* October 15

Summer Time in Great Britain and Northern Ireland ends on October 24.

MERCURY remains unobservable to observers in the latitudes of the British Isles though further south it is visible as an evening object throughout the month. For observers in southern latitudes this will be the most favourable evening apparition of the year. Figure 11 shows, for observers in latitude S.35°, the changes in azimuth (true bearing from the north through east, south, and west) and altitude of Mercury on successive evenings when the Sun is 6° below the horizon. This condition is known as the end of evening civil twilight, and in this latitude and at this time of year occurs about 30 minutes after sunset. The changes in the brightness of the planet are indicated by the relative sizes of the circles marking Mercury's position at five-day intervals. It will be noticed that Mercury is at its brightest before it reaches greatest eastern elongation (25°) on October 14. Mercury's magnitude is −0.1 at the beginning of the month, falling to +1.5 at the end.

VENUS, magnitude −3.8, continues to be visible as a brilliant morning object in the eastern sky before dawn. It is gradually drawing closer to the Sun, being just under 20° away by the end of the month.

MARS becomes lost to view over the western horizon as darkness falls.

JUPITER passes through conjunction on October 18 and therefore remains too close to the Sun for observation throughout the month.

SATURN, magnitude +0.5, continues to be visible as an evening object. Its slow retrograde motion has taken it back into Capricornus, reaching its second stationary point on October 28.

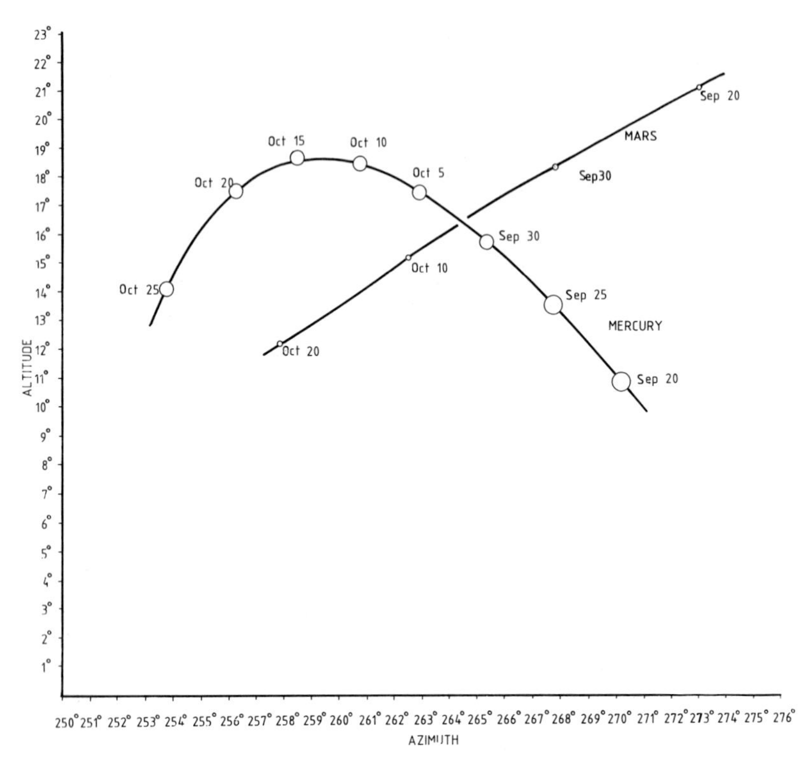

Figure 11. The evening apparitions of Mercury for latitude S.35°.

MIGRATING STARS. During summer evenings, and well into early autumn, the evening sky is dominated by three bright stars – Vega in Lyra, Altair in Aquila, and Deneb in Cygnus – which form a large triangle. Years ago, the Editor of this *Yearbook*, in a 'Sky at Night' television programme, nicknamed this arrangement the Summer Triangle. To his surprise the name has come into general use, even though it is completely unofficial, because the three stars are in different constellations, and in any case it is inapplicable to the Southern Hemisphere, where it really should be re-named the Winter Triangle.

Aquila, the Eagle, is a well-formed constellation – Altair is flanked on either side by two fainter stars, Gamma Aquilæ (Tarazed), which is orange, and the white Beta Aquilæ (Alshain) – yet a strange thing happened to it in December 1992. It lost one of its stars, Rho Aquilæ, which has migrated across the border of the adjacent constellation, Delphinus (the Dolphin).

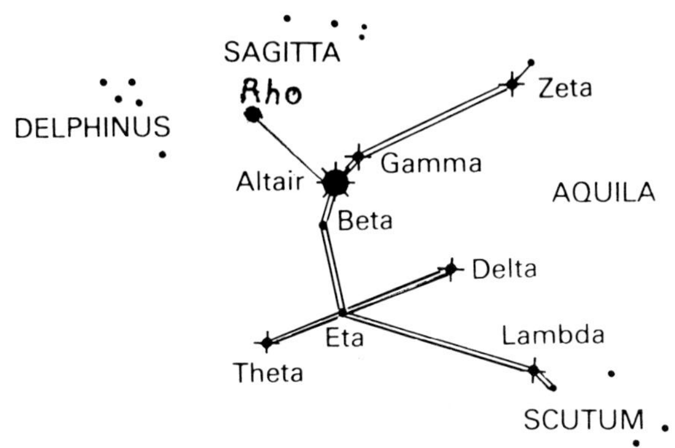

Figure 12. Migrating stars.

Stellar proper motions are very slight, and changing constellations is rare. Moreover, the constellations themselves are quite arbitrary, since the stars in them are not associated with each other. Our system is based on that of the Greeks, though it has been drastically modified, and the present boundaries have been laid down by the International Astronomical Union. Rho Aquilæ has been right on the border between the Eagle and the Dolphin; it has now 'crossed the floor'.

It is interesting to look at other cases of naked-eye stars which will also change constellations in the foreseeable future:

In the year	*Star*		*Constellation*
2400	Gamma Cæli		Columba
2640	Epsilon Indi		Tucana
2920	Epsilon Sculptoris		Fornax
3200	Lambda Hydri	*will*	Tucana
4500	Mu[1] Cygni	*move*	Pegasus
5200	Chi Pegasi	*into*	Pisces
5200	Mu Cassiopeiæ		Perseus
6300	Eta Sagittarii		Corona Australis
6400	Zeta Doradûs		Pictor

Obviously, these migrations are of academic interest only – and who can tell whether the constellations in use in the year 6400 will be the same as those which we use today?

PLANETS IN RETROGRADE MOTION. Planets, of course, change constellations very quickly, though with the exception of Pluto (which is probably unworthy of true planetary status) they keep strictly to the Zodiac, which stretches right round the sky. However, the superior planets (that is to say, those beyond the orbit of the Earth) do not move regularly from west to east against the starry background. As the Earth 'catches them up', so to speak, and passes them, the planet will stop, move backwards or retrograde for a while – east to west – and stop once more before resuming its usual direction of motion.

This is how Saturn is behaving at the present time. It has been in Aquarius, but its retrograde movement has now carried it back into Capricornus, where it will remain for most the rest of the year. It reaches its second stationary point at the end of October, and then begins to move eastward again, but it will be the very end of 1993 before it returns to the boundary between Capricornus and Aquarius.

LIMITS OF NAKED-EYE VISIBILITY. It is usually said that the faintest stars normally visible with the naked eye, under good conditions, are of magnitude 6. In general this may be true, but certainly there are some lynx-eyed people who can see stars well below this limit – particularly from countries where there is little light pollution.

A good test can be made this month. The main constellation of northern autumn or southern spring is Pegasus, the Flying Horse, whose four chief stars make up a square (though for quite unknown reasons one of the four, Alpheratz, has been transferred to Andromeda as Alpha Andromedæ; it used to be Delta Pegasi). It is interesting to see how many stars can be seen within the Square on a really dark, transparent night. If any readers of this *Yearbook* care to make the attempt, I will be interested to know – and if enough replies are received, I will give the results in our 1994 edition!

NOVEMBER

Full Moon: November 29 *New Moon:* November 13

MERCURY is a difficult morning object for the first few days of the month, for observers in equatorial and southern latitudes. It passes through inferior conjunction on November 6, when it transits the Sun (see below), and within a week re-appears as a morning object. For observers in northern temperate latitudes this will be the most favourable morning apparition of the year. Figure 13 shows, for observers in latitude N.52°, the changes in azimuth (true bearing from the north through east, south, and west) and altitude of Mercury on successive evenings when the Sun is 6° below the horizon. This condition is known as the beginning of morning civil twilight, and in this latitude and at this time of year occurs about 35 minutes before sunrise. The changes in the brightness of the planet are indicated by the relative sizes of the circles marking Mercury's position at five-day intervals. It will be noticed that Mercury is at its brightest after it reaches greatest western elongation (20°) on November 22. During the second half of November Mercury's magnitude brightens from +0.7 to −0.6. The transit of Mercury, which is a near-grazing one, will be visible from the Pacific Ocean, Antarctica, Australasia, Asia, eastern and southern Africa, and eastern Europe. It starts at $03^h\ 06^m$ and ends at $04^h\ 47^m$ G.M.T.

VENUS is a brilliant morning object with a magnitude of −3.8. However, it is moving closer in towards the Sun so that the period available for observation is shortening noticeably from week to week and by the end of the month it is only visible for a very short while after sunset. Reference to Figure 13 shows that Venus and Mercury are only a few degrees apart during the second half of November.

MARS is unsuitably placed for observation.

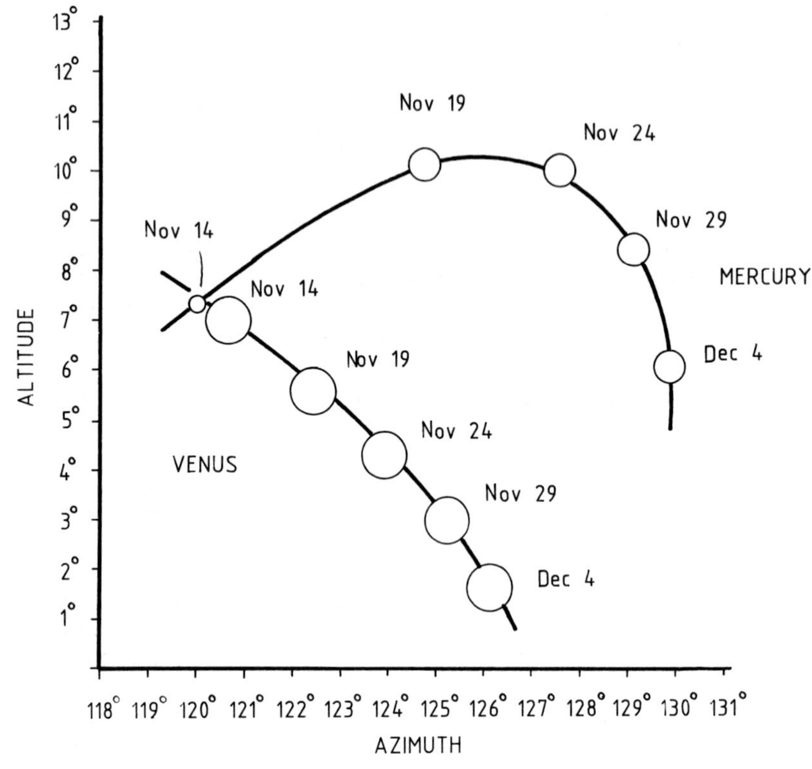

Figure 13. The morning apparition of Mercury for latitude N.52°.

JUPITER becomes visible as a bright morning object after about the first week of the month, magnitude −1.7. It may be detected low in the south-eastern sky for a short while before dawn. Jupiter and Venus will be seen close together in the sky around November 8–9, Venus being 2 magnitudes brighter than Jupiter.

SATURN, magnitude +0.7, continues to be visible as an evening object in Capricornus.

THE LEONID METEORS. The November Leonids, associated with the periodical comet Tempel–Tuttle, reach their maximum on the 17th of the month. Normally the ZHR or Zenithal Hourly Rate is very low, but there are occasional 'storms', as for instance those of 1799,

1833, 1866, and 1966. The average interval between these 'storms' is 33 years, but 1899 and 1933 were missed, because the orbit of the densest part of the swarm was perturbed by Jupiter and Saturn.

Things are much more encouraging for 1999, and indeed it has been predicted, with fair confidence, that there will also be a rich shower in 1998. This lies some time ahead yet, and we do not really hope for anything spectacular in 1993; but with the Leonids, one never knows, and it is certainly worth keeping a watch. The great 'storms' are short-lived. At the 1966 display about 60,000 meteors were seen over a period of forty minutes; but the peak occurred during daylight in Europe, and so was not seen. At any rate, meteor observers will be on 'Leonid alert' from now until the end of the century.

The associated comet is named after its discoverers in 1865–6; Wilhelm Tempel on December 19, 1865, and Horace Tuttle on the following January 6. The magnitude was then 3. The comet proved to have a period of just over 33 years. It was missed in 1899 and 1933, but it was recovered in June 1965 by Schubart, who detected it on photographs taken by M. Bester at the Boyden Observatory in South Africa. It was, however, very faint, and never exceeded magnitude 16; neither did it develop a tail. However, there seems little doubt that it is identical with comets seen in 1366 (maximum magnitude 3) and in 1699 (magnitude 4). How it will behave at the coming return remains to be seen.

LINNÉ AND LUNAR ECLIPSES. This month's total lunar eclipse should be well seen over much of Europe – clouds permitting. It cannot be said that an eclipse of the Moon is anything like so spectacular or so important as an eclipse of the Sun, but it is certainly worth watching, and interesting observations can be made.

As soon as the supply of direct sunlight is cut off, the temperature on the lunar surface falls sharply. In the past it has been claimd that there are perceptible effects on various specific features, and attention has been concentrated upon Linné, in the Mare Serenitatis or Sea of Serenity.

Beer and Mädler, the first major lunar cartographers, published a map in 1838 together with a full description of the surface. Linné was described as a deep, well-formed though small crater, not too unlike Bessel, the largest crater on the Mare Serenitatis. In 1866 Julius Schmidt, observing from Athens, announced that Linné no

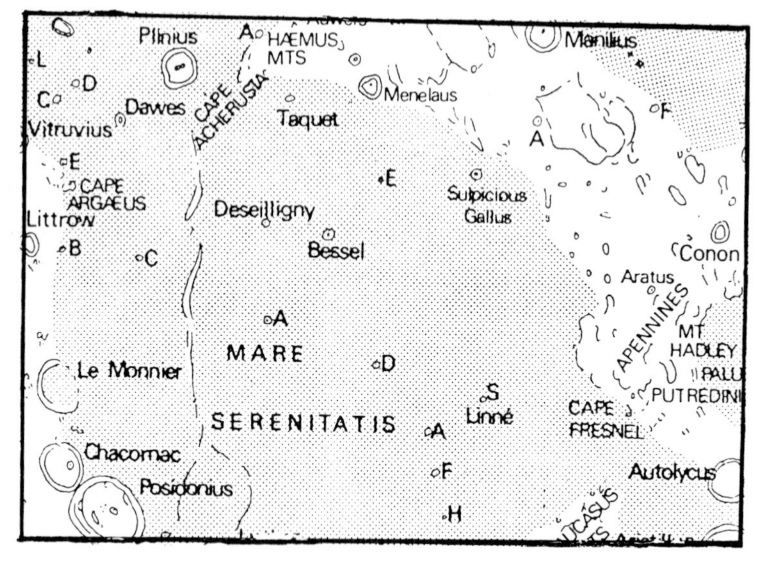

Figure 14. Map of the Mare Serenitatis showing Linné.

longer answered to this description; it had become a white patch. Subsequently it was found that Linné is a craterlet surrounded by a bright nimbus, and space-craft pictures have shown that it is a normal bowl-shaped craterlet, almost certainly due to impact. There seems no chance that any real change occurred between 1838 and 1866 (particularly as Mädler himself commented that he had observed Linné on both dates, and found no alteration at all), but the whole episode caused great discussion, and one American astronomer, W. H. Pickering, maintained that when Linné is suddenly cooled during an eclipse the bright nimbus increases in size and brightness – presumably because of some sort of deposit.

This too seems excessively unlikely, if only because the Moon is to all intents and purposes devoid of atmosphere. Yet strange optical effects can occur, and it is worth checking on Linné during the present eclipse.

JEAN SYLVAN BAILLY. This month is the second centenary of the death of a great French astronomer, Jean Sylvan Bailly. He was born in Paris on September 15, 1736, and became well known for his work on comets and stellar positions. He also wrote several notable books, such as his *Histoire d'Astronomie*. He was universally respected in scientific circles, but he was also prominent on the popular side during the French Revolution, and became Mayor of Paris – with the sad but predictable result that he was arrested, charged with conspiracy, and guillotined. He died on November 21, 1793 – one of a fortunately small band of astronomical martyrs.

Bailly is commemorated by a vast walled plain near the edge of the visible surface of the Moon. It is well over 180 miles in diameter, and though it has been described as 'a field of ruins' it is truly imposing when seen under suitable conditions of libration and illumination.

DECEMBER

Full Moon: December 28 *New Moon:* December 13

Solstice: December 21

MERCURY continues its morning apparition, with a magnitude of −0.6. Although observers in the British Isles are unlikely to see it after about December 5, those further south can hope to see it until the middle of the month, low above the south-eastern horizon until twilight inhibits observation.

VENUS, magnitude −3.8, is a bright morning object for the first week of December though it is then only visible low above the south-eastern horizon for a short while before sunrise. Thereafter it remains too close to the Sun for observation.

MARS is too close to the Sun for observation, passing through conjunction on December 27.

JUPITER, magnitude −1.8, is a bright morning object in the south-eastern sky before twilight inhibits observation.

SATURN is still visible as an evening object in the south-western sky, magnitude +0.8. It ends the year on the border between the constellations of Capricornus and Aquarius. Observers in the British Isles will find that by the end of December it is too low for observation after about 19^h.

PHÆTHON: ASTEROID OR DEAD COMET? In 1983 J. Davis and S. Green were examining data supplied by IRAS, the Infra-Red Astronomical Satellite, when they discovered a remarkable asteroid, now numbered 3200 and named Phæthon after the mythological boy who persuaded his father Helios to let him drive the Sun-chariot, with disastrous results. Phæthon is not an 'Earth-grazer', but its orbit takes it well within that of Mercury; its distance from the Sun ranges between 21,000,000 km (13,000,000 miles) and 390,000,000 km (242,000,000 miles), so that at aphelion it is well into the main asteroid zone. It and 1566 Icarus are the only asteroids known to invade these torrid parts of the Solar System.

Phæthon is about three miles in diameter, and has a rotation period of approximately four hours. It seems to be darkish. However, the real interest stems from the fact that its orbit is very like that of the Geminid meteor stream. The Geminids produce good displays every year between December 7 and 16, with their peak on the 13th; the ZHR can sometimes be as high as 75, so that they can rank with the August Perseids. No known comet is associated with them – so can they be debris left by Phæthon in its former cometary guise?

Whether or not small, irregular asteroids are extinct comets has been under discussion for many years now. The dimensions are suitable; even a great comet, such as Halley's, has a small nucleus. Yet serious doubts remain, and some astronomers are openly sceptical.

There have even been suggestions that a very different sort of object – Chiron – may be re-classified as a comet! Chiron, discovered by C. Kowal in 1977, spends almost all its time moving between the orbits of Saturn and Uranus, far beyond the main belt of asteroids; it is large by asteroidal standards, and is certainly at least 100 miles in diameter. It is next due at perihelion in 1995, and has recently developed what appears to be a coma. This is cometary behaviour, but Chiron seems to be far too large to be classed as such. Its nature is still uncertain; Kowal, the discoverer, summed the situation up neatly when he said that Chiron was – 'well, just Chiron!'

THE RETURN OF ORION. Orion, the Hunter, has returned to the evening sky by December, and will dominate the scene from now until the middle of northern spring or southern autumn. With its two brilliant leaders – the glittering white Rigel and the red supergiant Betelgeux – it is unmistakable; there is also the shining Belt and the misty Sword, marked by the Great Nebula, M.42, which is over 1000 light-years away, and is a splendid 'Nursery' in which fresh stars are being formed from the nebular material.

Orion is cut by the celestial equator, which passes very close to Mintaka or Dekta Orionis, the faintest of the three Belt stars. This means that the Hunter is visible from every inhabited part of the

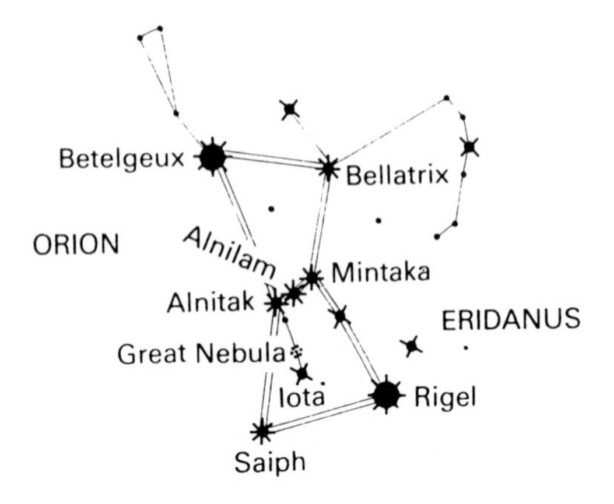

Figure 15. Orion the Hunter.

world. Indeed, there is no site from which at least some of it cannot be seen. From the North Pole, Betelgeux is always above the horizon, while Rigel never rises; from the South Pole the converse is true. From the Equator, Orion passes across the zenith. It is fascinating to see the Hunter overhead, with the Great Bear on one side of the sky and the Southern Cross on the other!

Eclipses in 1993

In 1993 there will be four eclipses, two of the Sun and two of the Moon.

1. *A partial eclipse of the Sun on May 21* is visible from North America except the south-east, Arctic regions, Greenland, Iceland, northern Europe including the north of the British Isles, and north-west Russia. The eclipse begins at $12^h\,9^m$ and ends at $16^h\,20^m$. At the time of maximum eclipse 0.74 of the Sun's diameter is obscured.

2. *A total eclipse of the Moon on June 4* is visible from the extreme tip of South America, the western coast of North America, the Pacific Ocean, Antarctica, Australasia and south-east Asia. The eclipse begins at $11^h\,12^m$ and ends at $14^h\,50^m$. Totality lasts from 12^h 13^m until $13^h\,49^m$.

3. *A partial eclipse of the Sun on November 13* is visible from the southern tip of South America, Antarctica, most of New Zealand and south Australia. The eclipse begins at $19^h\,46^m$ and ends at 23^h 43^m. At time of maximum eclipse 0.93 of the Sun's diameter is obscured.

4. *A total eclipse of the Moon on November 29* is visible from most of Europe including the British Isles, west Africa, Iceland, Greenland, the Arctic regions, the Americas and north-east Asia. The eclipse begins at $04^h\,40^m$ and ends at $08^h\,10^m$. Totality lasts from 06^h 02^m to $06^h\,48^m$.

Occultations in 1993

In the course of its journey round the sky each month, the Moon passes in front of all the stars in its path, and the timing of these occultations is useful in fixing the position and motion of the Moon. The Moon's orbit is tilted at more than five degrees to the ecliptic, but it is not fixed in space. It twists steadily westwards at a rate of about twenty degrees a year, a complete revolution taking 18.6 years, during which time all the stars that lie within about six and a half degrees of the ecliptic will be occulted. The occultations of any one star continue month after month until the Moon's path has twisted away from the star but only a few of these occultations will be visible at any one place in hours of darkness.

There are six occultations of bright planets in 1993, two of Mercury, three of Venus, and one of Mars. Several of them are visible from North America.

Only four first-magnitude stars are near enough to the ecliptic to be occulted by the Moon; these are Regulus, Aldebaran, Spica, and Antares. No occultations of these stars occur in 1993.

Predictions of these occultations are made on a world-wide basis for all stars down to magnitude 7.5, and sometimes even fainter. The British Astronomical Association has just produced the first complete lunar occultation prediction package for microcomputer users.

Recently occultations of stars by planets (including minor planets) and satellites have aroused considerable attention.

The exact timing of such events gives valuable information about positions, sizes, orbits, atmospheres and sometimes of the presence of satellites. The discovery of the rings of Uranus in 1977 was the unexpected result of the observations made of a predicted occultation of a faint star by Uranus. The duration of an occultation by a satellite or minor planet is quite small (usually of the order of a minute or less). If observations are made from a number of stations it is possible to deduce the size of the planet.

The observations need to be made either photoelectrically or visually. The high accuracy of the method can readily be appreciated when one realizes that even a stop-watch timing accurate to $0^s.1$ is, on average, equivalent to an accuracy of about 1 kilometre in the chord measured across the minor planet.

Comets in 1993

The appearance of a bright comet is a rare event which can never be predicted in advance, because this class of object travels round the Sun in an enormous orbit with a period which may well be many thousands of years. There are therefore no records of the previous appearances of these bodies, and we are unable to follow their wanderings through space.

Comets of short period, on the other hand, return at regular intervals, and attract a good deal of attention from astronomers. Unfortunately they are all faint objects, and are recovered and followed by photographic methods using large telescopes. Most of these short-period comets travel in orbits of small inclination which reach out to the orbit of Jupiter, and it is this planet which is mainly responsible for the severe perturbations which many of these comets undergo. Unlike the planets, comets may be seen in any part of the sky, but since their distances from the Earth are similar to those of the planets their apparent movements in the sky are also somewhat similar, and some of them may be followed for long periods of time.

The following periodic comets are expected to return to perihelion in 1993, and to be brighter than magnitude +15.

Comet	*Year of discovery*	*Period (years)*	*Predicted date of perihelion 1993*
Ciffreo	1985	7.2	Jan. 22
Schaumasse	1911	8.2	Mar. 4
Ashbrook-Jackson	1948	7.5	July 14
West-Kohoutek-Ikemura	1974	6.4	Dec. 25

Minor Planets in 1993

Although many thousands of minor planets (asteroids) are known to exist, only 3,000 of these have well-determined orbits and are listed in the catalogues. Most of these orbits lie entirely between the orbits of Mars and Jupiter. All of these bodies are quite small, and even the largest, Ceres, is only about 960 kilometres in diameter. Thus, they are necessarily faint objects, and although a number of them are within the reach of a small telescope few of them ever reach any considerable brightness. The first four that were discovered are named Ceres, Pallas, Juno and Vesta. Actually the largest four minor planets are Ceres, Pallas, Vesta and Hygeia (excluding 2060 Chiron, which orbits mainly between the paths of Saturn and Uranus, and whose nature is uncertain). Vesta can occasionally be seen with the naked eye and this is most likely to occur when an opposition occurs near June, since Vesta would then be at perihelion. Approximate dates of opposition (and magnitude) for these minor planets in 1993 are: Ceres, October 22 ($7^m.4$), Pallas, August 24 ($9^m.1$), and Vesta, August 28 ($5^m.9$).

A vigorous campaign for observing the occultations of stars by the minor planets has produced improved values for the dimensions of some of them, as well as the suggestion that some of these planets may be accompanied by satellites. Many of these observations have been made photoelectrically. However, amateur observers have found renewed interest in the minor planets since it has been shown that their visual timings of an occultation of a star by a minor planet are accurate enough to lead to reliable determinations of diameter. As a consequence many groups of observers all over the world are now organizing themselves for expeditions should the predicted track of such an occultation cross their country.

In 1984 the British Astronomical Association formed a special Asteroid and Remote Planets Section.

Meteors in 1993

Meteors ('shooting stars') may be seen on any clear moonless night, but on certain nights of the year their number increases noticeably. This occurs when the Earth chances to intersect a concentration of meteoric dust moving in an orbit around the Sun. If the dust is well spread out in space, the resulting shower of meteors may last for several days. The word 'shower' must not be misinterpreted – only on very rare occasions have the meteors been so numerous as to resemble snowflakes falling.

If the meteor tracks are marked on a star map and traced backwards, a number of them will be found to intersect in a point (or a small area of the sky) which marks the radiant of the shower. This gives the direction from which the meteors have come.

The following table gives some of the more easily observed showers with their radiants; interference by moonlight is shown by the letter M.

Limiting dates	*Shower*	*Maximum*	*R.A.*	*Dec.*	
Jan. 1–4	Quadrantids	Jan. 4	15^h28^m	+50°	M
April 20–22	Lyrids	April 21	18^h08^m	+32°	
May 1–8	Aquarids	May 5	22^h20^m	+00°	M
June 17–26	Ophiuchids	June 20	17^h20^m	−20°	
July 15–Aug. 15	Delta Aquarids	July 29	22^h36^m	−17°	M
July 15–Aug. 20	Piscis Australids	July 31	22^h40^m	−30°	M
July 15–Aug. 25	Capricornids	Aug. 2	20^h36^m	−10°	M
July 27–Aug. 17	Perseids	Aug. 12	3^h04^m	+58°	
Oct. 15–25	Orionids	Oct. 22	6^h24^m	+15°	
Oct. 26–Nov. 16	Taurids	Nov. 3	3^h44^m	+14°	M
Nov. 15–19	Leonids	Nov. 17	10^h08^m	+22°	
Dec. 9–14	Geminids	Dec. 13	7^h28^m	+32°	
Dec. 17–24	Ursids	Dec. 23	14^h28^m	+78°	M

M = moonlight interferes

Some Events in 1994

ECLIPSES

There will be three eclipses, two of the Sun and one of the Moon.

May 10: annular eclipse of the Sun – Americas, Europe.
May 25: partial eclipse of the Moon – Africa, Europe, Americas.
November 3: total eclipse of the Sun – South America, Southern Africa.

THE PLANETS

Mercury may be seen more easily from northern latitudes in the evenings about the time of greatest eastern elongation (February 4) and in the mornings around greatest western elongation (November 6). In the Southern Hemisphere the dates are March 19 (morning) and September 26 (evening).

Venus is visible in the evenings from late February to October and in the mornings from November onwards.

Mars is visible in the mornings from March onwards.

Jupiter is at opposition on April 30.

Saturn is at opposition on September 1.

Uranus is at opposition on July 17.

Neptune is at opposition on July 14.

Pluto is at opposition on May 17.

The Invention of the Reflecting Telescope

COLIN RONAN

To most people the origin of the reflecting telescope is associated with Isaac Newton. Dissatisfied with the coloured fringes displayed by images in the refractor, Newton had by 1669 built a reflecting telescope with a mirror made of speculum metal – an amalgam of copper, tin and a little antimony. But Newton did not claim to have invented the principle of the reflector. Indeed, he had already examined a theoretical design by the Scots mathematician James Gregory, which made use of a concave mirror for gathering light – as Newton himself was to use – and a small concave elliptical mirror to intercept the focused light beam and bring the rays through a hole in the larger concave mirror to an eyepiece directly behind it. Moreover, designs had also been proposed by Marin Mersenne and a Monsieur Cassegrain in France, and by Niccolo Zucchi and Bonaventura Cavalieri in Italy. Clearly, then, the origin of that prime tool of astronomers, the reflector, must almost certainly be earlier than any of these – and that takes us back before 1632.

Studies of the reflection of light by mirrors goes back a very long way before this. The Greek mathematician Euclid had written on the subject as early as about 300 BC, explaining the use of concave mirrors to bring sunlight to a focus, and certainly such 'burning mirrors' became well known in later times. However, it may well be that they were known before Euclid wrote about them. Nevertheless, it seems to have been only in the thirteenth century that the subject was pursued again with any vigour. Then Robert Grosseteste, a central figure in English intellectual life at Oxford, wrote a series of scientific books between 1220 and 1235. These included a number of treatises on light and optics, and in one of them he refers to making '. . . things a very long distance off appear as if very close . . .' and doing so by making the light penetrate '. . . through several transparent media of different natures'. But though this may be a description of the principle of the refracting telescope, Grosseteste gives no diagram to make his suggestion more specific.

Grosseteste's most notable pupil was Roger Bacon, who some thirty years later wrote again in similar terms, and discussed reflection by mirrors. In the early 1920s some interest was shown in a manuscript diary attributed to Roger Bacon and written in code. The code was deciphered by Romaine Newbold of Pennsylvania University, and it seems that the diary contained some astronomical observations which, it was claimed, could only have been made using a telescope. However, astronomers and historians of science have remained totally unconvinced of the validity of this claim.

Bacon was followed by two other thirteenth-century optical writers, Witelo from eastern Europe and the Englishman John Pecham. Both considered in detail reflection from curved as well as flat surfaces, Pecham in particular making a careful analysis of image formation. After this the idea of seeing things at a distance was current among some scientifically minded men, not least Geoffrey Chaucer, the civil servant and diplomat, who wrote and published two treatises on the astrolabe and its use in astronomy, as well as his better-known *Canterbury Tales* written about 1386. Here, in 'The Squire's Tale', he describes the use of such an instrument, but it is used for detecting distant scenes of marital infidelity, not observing the heavens.

The work of Leonard Digges

Leonard Digges was born sometime about 1520, and it is only with him that the reflecting telescope appears to have become more than a dream; at least there seem to be no separately attested records of any others exploring the matter in a practical way before his time. Indeed, this is the crucial factor in discovering the beginnings of the reflector. There is no independent evidence of Roger Bacon's telescope – if he ever really did have one – but for Digges matters are different, as will become evident shortly.

Leonard was a gentleman of private means, and was interested in the practical applications of mathematics – gunnery, surveying and navigation – and he produced almanacs and prognostications (forecasts) of use to mariners. According to his son, Thomas, he also carried out practical as well as theoretical work on both the reflector and the refractor. That this is true, also has independent confirmation from a close friend of Leonard Digges, the mathematician John Dee who became an adviser to Queen Elizabeth I. Interested too in navigational problems, and at a time when the human life span was often short, Dee agreed to arrangements proposed by Leonard

Digges that, in the event of Digges' early death, Dee should act as guardian of his son Thomas. In the event this proved to be a very wise decision.

As a member of a noted Kent family, Leonard Digges took part in the unsuccessful rebellion of the men of Kent led by Sir Thomas Wyatt. This rebellion, against the marriage of Queen Mary with Prince Philip of Spain, led to Wyatt being beheaded in April 1554, and very nearly to the execution of Leonard Digges. Indeed, Digges was saved only by a reprieve, granted probably due to the intervention a kinsman, Lord Clinton, to whom he had dedicated one of his almanacs. Nevertheless, all Digges' property was confiscated and he had to spend the next four years completing payments for its redemption. Once he had done this, Leonard Digges, still a comparatively young man of only about thirty-nine years of age, died.

Dee took Thomas, then aged about thirteen, under his wing and saw to his education. Dee lived at Mortlake, south of London, and possessed a large library which contained an extensive collection of scientific books including a considerable number of manuscripts of Roger Bacon. Unfortunately these were all later destroyed when the library was burned to the ground by a mob in 1583, but during Thomas's formative years, they were there to be consulted. So Thomas would have become well aware of Roger Bacon's work, and that of Witelo and Pecham as well.

The claims of Thomas Digges

Thomas Digges, whose date of birth is uncertain but was probably 1546, followed in his father's footsteps, becoming an expert surveyor. Then in 1572, at the age of about twenty-six, he sat for parliament, representing Wallingford and later, in the parliament of 1585, he represented Southampton. The next year Thomas was appointed 'muster-master' general to the British forces in Holland, which were helping the Netherlanders' fight for their independence against King Philip of Spain. This last probably brought him into contact with his opposite number in the Netherlands' forces. This is significant because it may well be that he discussed the use of the refracting telescope, with which his father Leonard had also experimented, with a senior military surveyor, Adriaen Anthonisz (also known as Adriaen Metius), whose son Jacob Metius was later to lay claim to being the inventor of the instrument. These political events are important in another way, as we shall now see.

By 1570 Thomas's guardian, John Dee, was promoting the use of telescopes by military commanders. (They were, however, not called 'telescopes' but often referred to as 'perspective glasses'.) This arose because in that year Henry Billingsley published the first edition in English of Euclid's famous and fundamental treatise on geometry known as the *Elements*, and Dee wrote a famous preface to the work. In this Dee not only supported the new Copernican theory, published only twenty-seven years earlier, but also discussed the use of geometry in military tactics, and said that a commander 'may wonderfully help him selfe, by perspective Glasses. In which (I trust) our posterity will prove more skillful and expert, and to greater purposes, then in these dayes, can (almost) be credited to be possible.'

Then the very next year, 1571, Thomas Digges published a book on surveying and geometry. Known usually as *Pantometria*, the book had a long explanatory title which is worth quoting in full. It runs: *A Geometrical Practise, named Pantometria, divided into three Bookes, Longimetra, Planimetra and Stereometria, containing Rules manifolde for mensuration of all lines, Superficies and Solides: with sundry strange conclusions both by instrument and without, and also by Perspective glasses, to set forth the true description or exact plat of an whole Region: framed by Leonard Digges, Gentleman, lately finished by Thomas Digges, his sonne*. The Perspective glasses to which he referred were devices for seeing at a distance as the text indicates, though in practical terms he only discusses in detail the use of a plane mirror in surveying, in this case determining the distances from land of ships at sea. Yet he is quite clear about what his father Leonard had achieved. He wrote:

> '. . . my father by his continual painfull practises [i.e. practical experiments], assisted with Demonstrations *Mathematicall*, was able, and sundrie Times hath by proportionall Glasses duly situate in convenient angles, not onely discovered things farre off, read letters, numbered peeces of money with the very coyne and superscription thereof, cast by some of his freends uppon Downes in open fieldes, but also seven myles of declared what hath been doon at that instant in private places . . .'

and a few pages later, more significantly still:

> 'Thus much I thought good to open concerning the effects of a plaine Glasse [i.e. a plane mirror], very pleasant to practise, yea

> most exactly serving for the description of a plaine champion country. But marvellous are the conclusions that may be performed by glasses concave and convex of Circulare and parabolicall formes, using for multiplication of beames sometimes the aide of Glasses transparent, which by fraction [i.e. refraction] should unite or dissipate the images or figures presented by the reflection of the other. By these kinds of Glasses or rather frames of them, placed in due Angles, yee may not only set out before your eye the littely image of every Towne, Village, etc. and that in as little or great space or place as ye will prescribe, but also augment and dilate any parcell thereof . . ., ye may by application of Glasses in due proportion cause any peculiare house, or roume thereof dilate and shew it selfe in as ample forme as the whole towne first appeared, so that ye shall discerne any trifle, or read any letter lying there open, especially if the the sunne beames come unto it, as plainly as if you were corporally present, although it be distant as from you as farre as the eye can discrye: . . .'

This, surely, is neither more nor less a description of a reflecting telescope and what can be seen through it at different magnifications. However, Digges has one final remark to add:

> 'But of these conclusions I minde not here more to intreate, having at large in a volume by it selfe opened the miraculous effects of perspective glasses.'

No copy of any such book by Thomas Digges has yet been found, and some modern historians have taken this to bolster a view that all he wrote in the *Pantometria* was no more than an imaginative description of what might be achieved. There never was such a book, they claim; Digges could never have written one, they say, because he had not the necessary knowledge. However, there is another explanation which seems far more probable in the light of what Digges certainly did write, and taking into account subsequent events, as we shall now see.

Independent evidence

Perhaps the most powerful argument against the view that Thomas Digges did not really know what he was writing about, lies in the evidence of an independent report of the 1570s written by

William Bourne. A practical mathematician like Digges and Dee, expert in navigation and nautical matters, he nevertheless had no inherited wealth and in consequence possessed no funds to enable him to experiment to any great extent. Thus he wrote:

> 'For that the hability of my purse ys not able for to reache, or beare the charges, for to seek thorowly what may be done with these two sortes of Glasses, that ys to say, the hollowe or concave glasse: and allso that glasse that ys grounde and polysshed rounde, and thickest in the middle, and thynnest towards the sides and edges.'

From what he says earlier in the document, Bourne's 'concave glasse' was without doubt a parabolic concave mirror, with its reflecting surface on the rear – 'the hylly side' – of the glass. He goes on:

> 'Therefore I can say the lesse unto that matter. For that there ys dyvers in this Lande that can say and dothe knowe muche more, in the causes than I: and specyally Mr Dee, and allso Mr Thomas Digges, for that by theyre Learning, they have reade and from many moo auctors in those causes: And allso their abillity ys suche, that they may the better mayntayne the charges [i.e. the experimental costs]. And allso they have more leysure and better tyme to practyze those matters, which ys not possible for mee, for to know in a nombre of causes, that thinge that they doo knowe. But notwithstanding upon the smalle prooffe and experyence that bee but unto small purpose, of the skylles and knowlledge of these causes, yet I am assured that the Glasse that ys grounde, beeynge of very clear stuffe, and of a good largenes, and placed so, that the beame doth come thorowe, and so reseaved into a very large concave looking Glasse, that it will shewe the thinge of marvelous largenes, in manner uncredable to bee beleeved of the common people.'

If we remember that in Bourne's time and long afterwards, down at least to the time of William Herschel a couple of centuries later, the largeness of lenses was a way of expressing their focal length, not their aperture. What Bourne writes in the last sentence just quoted is therefore immensely significant. In modern terms he says that if you take a lens of 'large' focal length and use it in conjunction

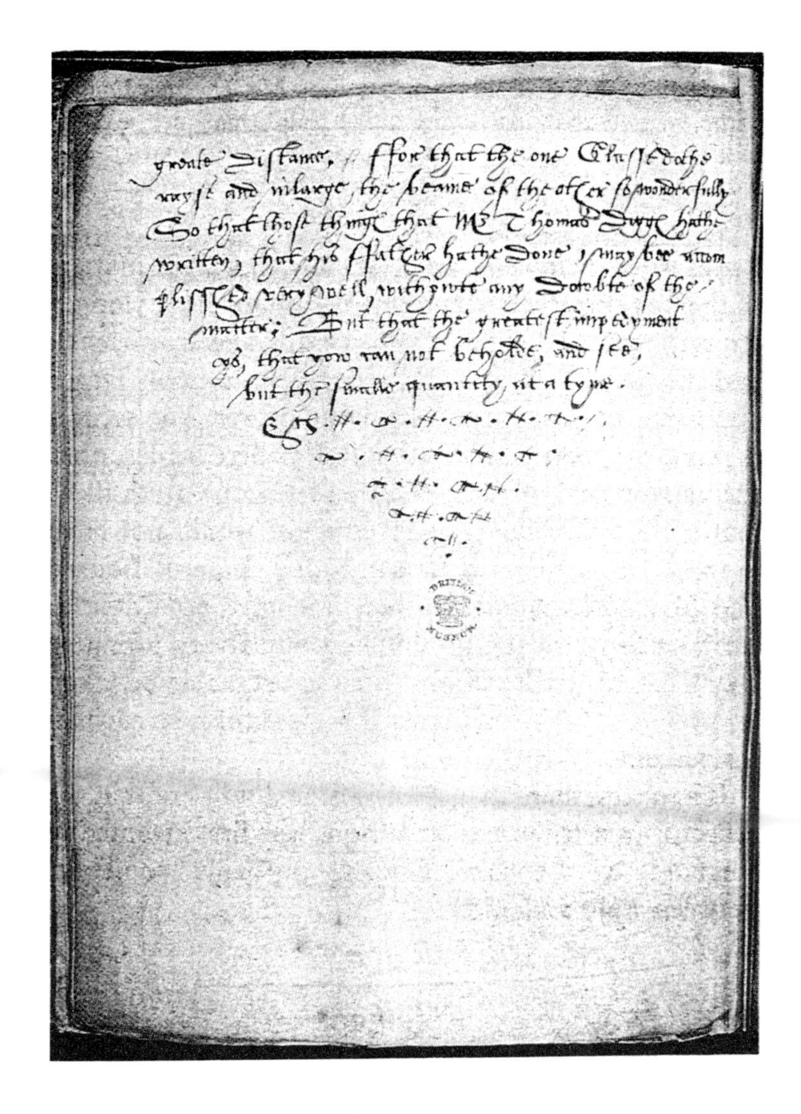

greate distance, ffor that the one Glasse dothe
rayse and inlarge, the beame of the other so wonderfully.
So that those thinges that Mr Thomas Digges hathe
written, that his ffather hathe done, may bee accom
plisshed very well, without any doubte of the
matter; But that the greatest impedyment
ys, that yow can not beholde, and see,
but the smaller quantity at a tyme.

Figure 1. The last and most crucial page of William Bourne's report to Lord Burghley which contains the reference to the small field of view of a telescope. This appears in the last sentence which reads: 'But that the greatest impediment ys, that yow cannot beholde, and see, but the smaller quantity at a time.' (Reproduced by permission of The British Library*)* with a mirror of even larger focal length, you get a good magnification. In fact what he is describing is a reflecting telescope.

But Bourne's text continues with an even more significant passage:

> 'Wherefore it is to be supposed, and allso I am of that opinyon, that having divers and sondry sortes of these concave looking glasses, made of great largeness, . . . yt ys likely yt ys true to see a smalle thinge, of great distance, ffor that the one Glasse dothe rayse and inlarge, the beame of the other so wonderfully. So that those things that Mr Thomas Digges hathe written that his father hathe done, may bee accomplisshed very well, withowte any dowbte of the matter. But that the greatest impediment ys, that yow cannot beholde, and see, but the smaller quantity at a tyme.'

Here Bourne first refers to the use of mirrors of longer focus for a greater magnification. Second, and even more significant, he points out that the instrument gives only a restricted field of view. It seems certain that Bourne himself could never have worked out these results theoretically; indeed it is unlikely that anyone else could do so at that time, at least as far as the size of the field of view is concerned. It is surely, therefore, evident that William Bourne had himself looked through a Digges reflector.

A matter of secrecy

William Bourne's report was a special commission. At the start of it he refers to a conversation with Lord Burghley, Queen Elizabeth I's chief adviser, who clearly commissioned the report. Moreover, in the whole of it, Bourne's style makes it clear that he was only willing to commit himself on what he had been able to confirm with his own eyes. It is indeed a good report, and one on which a senior government minister could rely.

The reason Lord Burghley should want such facts is not hard to find. Britain was in serious danger at this time. Intelligence reports made it clear that there was a scheme to overthrow Elizabeth's government, and that an invasion from Spain was imminent. These were no irrational fears. Bourne wrote his report in the 1570s; in 1588 the Spanish Armada set sail to conquer England. Burghley's commission was therefore a timely request from a far-sighted minister.

A 'Perspective Glasse' – a device for seeing at a distance – would be of inestimable use from military and naval points of view. It would enable a ship on the distant horizon to be identified unequivocally as friend or foe. Indeed, when twenty-three years later, in 1608, Hans Lipperhey asked the government of the Netherlands to grant him patents on the refracting telescope, that government wanted the whole thing kept secret, lest its enemies learn of the device.

In England such an outlook would clearly prevail as well, and with good reason. This, then, is probably why Thomas Digges' book explaining in detail his father's inventions never saw the light of day. Even a decade later, there was much spying and counter-espionage, and the government would have been very likely to request or order that it be not published. In brief, it seems not unrealistic to suggest that it suffered the Tudor equivalent of a 'D' notice.

Observational evidence

If, as William Bourne confirmed, the reflecting telescope of Digges was an actual fact, underlined by Bourne's description not only of how the concave mirror was made but the accompanying lens also, then we might expect that somewhere there is observational evidence of its use. This indeed appears to exist, not just in those descriptions already quoted, but in an astronomical context, as will now become evident.

In 1543 the famous book *De Revolutionibus Orbium Coelestium (On the Revolutions of the Celestial Orbs)* by Nicholas Copernicus was published. This placed the Sun, not the Earth, at the centre of the Universe. Certainly it led to a great leap in understanding planetary motion and, incidentally, in appreciating the place of humanity in the Universe, yet it still retained the ancient idea that the Universe was shaped like a sphere with the stars embedded on its inner surface. Admittedly observations showed that this sphere must be of very great size, but a sphere it was nevertheless. To Copernicus we still lived in a closed universe, bounded by an orb of stars.

As is well known, religious and philosophical arguments raged over the Copernican scheme, Galileo was barred from teaching it, and in Roman Catholic countries Copernicus's book was banned. Yet in Protestant England, things were freer. People could hold differing opinions, and could openly promote the Copernican theory. Dee did so in his Preface to Billingsley's book on Euclid's geometry, and his ward Thomas Digges made no bones about his

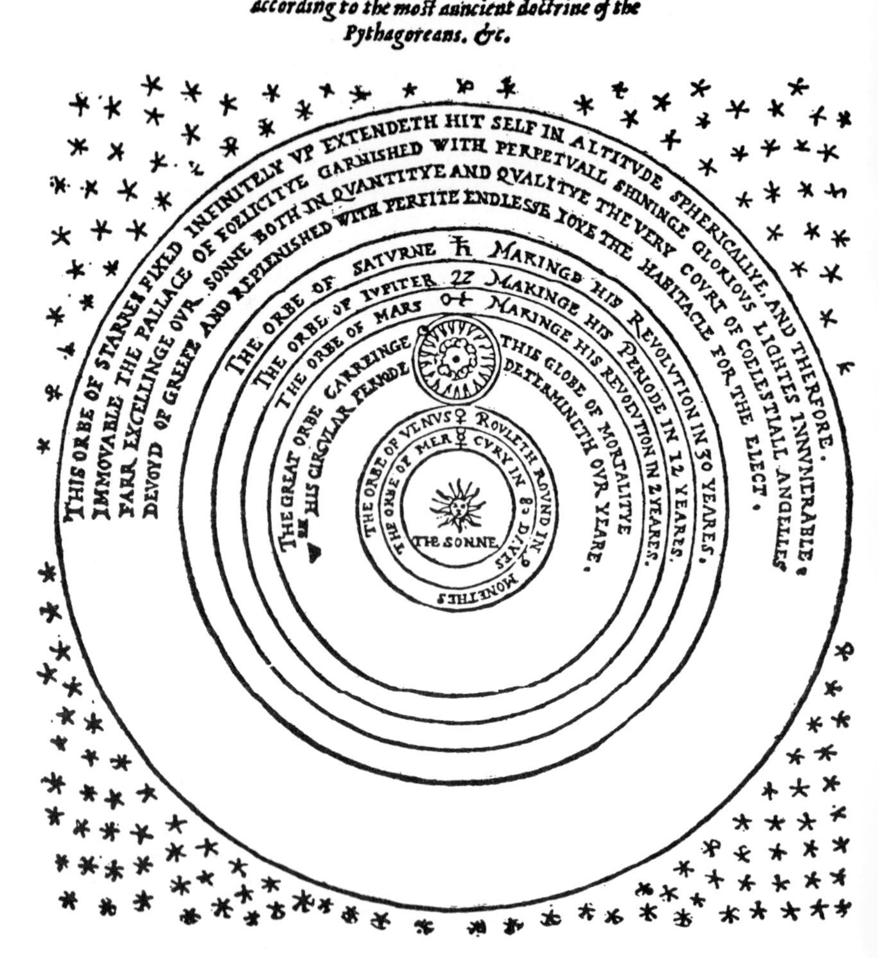

Figure 2. The first diagram to illustrate the proposal that the Universe is infinite. From the edition by Thomas Digges of his father's A Prognostication everlastinge . . ., *published in 1576 in London, eight years before its publication by Giordano Bruno to whom the idea is often credited. (Reproduced by permission of* The Royal Society*)*

support in a book he published in 1576. This was a new edition of his father's *Prognostication everlasting* . . ., consisting originally of a perpetual calendar with 'chosen rules to judge the weather by the Sunne, Moone, Starres, Comets, Rainebow, Thunder, Cloudes, . . .'. From our point of view, the significant thing is that Thomas Digges not only corrected and amended it, but also added an appendix which proves to be of singular importance.

With a title 'A Perfit Description of the Cælestiall Orbes according to the most ancient doctrine of the Pythagoreans, lately revived by Copernicus and by Geometricall Demonstrations approved', this sets out the Copernican system clearly and unequivocally. What is so significant, however, is that the text is accompanied by a fold-out plate, which is nothing short of astounding because it shows not a closed but an infinite universe. The outer sphere of stars is there but beyond it, as the diagram points out in Digges' own words 'This orbe of starres fixed infinitely up extendeth hit self in altitude spherically and therefore immovable, the palace of felicitye garnished with perpetuall shininge glorious lightes innumerable farr excelling our sonne both in quantitye and qualitye . . .'

This is a totally new idea. Admittedly, Digges might have felt that now the Earth was a planet orbiting the Sun, the crystal spheres of ancient astronomers would no longer be valid, and so the final sphere of the fixed stars should go too, but his description takes us further than this. His new universe is populated with innumerable stars extending outwards infinitely into space, some brighter than our Sun. This is an impression one could only get, surely, by looking through a telescope and seeing the myriad stars which then come into view.

Moreover, there is an additional factor. In an earlier part of his report to Lord Burghley, William Bourne is clear that the mirrors of reflecting telescopes he had seen had their reflecting surface on the rear side of the glass, not on the parabolic surface as is the practice now. On a dark sky, therefore, there would be ghost images, at least of the brighter stars, so in observing the sky Digges would have seen not only a myriad of stars, as Galileo was to do some thirty years later, but many more due to ghosting. This would indeed fit Digges's description perfectly. Moreover, a comment in the text to accompany the appendix makes this concept of infinity evident indeed, while what is said surely is a description of what an observer has seen, not what a theorist of the sixteenth century would describe. Digges writes:

'. . . that fixed Orbe garnished with lightes innumerable and reaching up in *Sphaericall* altitude without ende. Of which lightes Celestiall it is to be thoughte that we only beholde such as are in the inferioure partes of the same Orbe, and as they are hygher, so seeme they of lesser and lesser quantity, even till our sighte being not able farder to reache or conceyve, the greatest part reste by reason of their wonderful distance invisible to us.'

Conclusion

With the background to the Digges family and their undoubted integrity, the published evidence and claims of Thomas Digges in *Pantometria*, the totally independent and careful evidence of William Bourne, and the diagram and text in Thomas Digges' *Prognostication everlasting*, the evidence seems overwhelming. It becomes evident that Leonard Digges did indeed design and build the first reflecting telescope sometime between, say, 1545 and 1559. This is half a century before Dutch claims for a refracting telescope were made, and over a century before Isaac Newton constructed his little reflector.

IRIS: A Dream Fulfilled

DAVID ALLEN

Where shall I begin? Shall I tell of our astonishment when Michael, Vikki and I first aimed her at the weird galaxy Centaurus A? Should I try to recount the thrill when Jason and I saw the first spectrum of the supernova drawn on the screen? Shall I narrate the wild thoughts that raced through my mind when my analysis of the mass-loss star revealed that unexpected nebula? Or might I strive to capture the excitement Dave and I felt as, just moments before the planet set, we discovered the oxygen airglow?

Much though these memories jostle for attention as I sit at my word processor, before I give them air I must delve back several years to develop an introduction to these fun bits.

California, 1986

Dave Crisp enters my story as long ago as 1986. It was our first meeting. We travelled together through the car-choked freeways of Los Angeles, I on my way to visit one of the manufacturers of infrared detectors; he taking the opportunity to educate me on some of the subtleties of planetary atmospheres. From my perspective there was no subtlety to the chemical horrors of the LA atmosphere that day; but Dave's interests lay on another planet.

That week I visited both the companies that were developing two-dimensional infrared detectors. In their hands lay the key to one of the great strides so many of us foresaw and had long awaited. When working beyond the visible range, at infrared wavelengths, we had for too long had to grope around the sky with single detectors, blind outside each tiny spot where we chose to aim our telescopes. What other features the sky contained, what discoveries lay just outside our grasp, we could not guess. Now our myopia was to be removed by a new technology, by devices that could bring full vision to infrared astronomy.

There were pros and cons to both detectors. For use on the Anglo-Australian Telescope, however, I saw a clear preference for the material known as mercadtel, being manufactured by the Rockwell Science Center in the pleasant town of Thousand Oaks, west of Los Angeles. But in 1986 the rival material was considerably advanced, and the type of detector I preferred simply was not available. I opted to play a waiting game.

Hawaii 1987

To the backdrop of a continuous eruption from the flanks of Kilauea volcano, infrared astronomers convened six months later in Hilo on the island of Hawaii. Over the several days of the conference manufacturers released their latest performance figures and astronomers gloated over what they had achieved, or what they hoped to achieve, with these new devices.

It was an invigorating conference, made the more memorable when the Mayor of Hawaii dropped in, summoned me and Ben Zuckerman up on stage, and presented us with beautiful certificates (Figure 1) welcoming us to the island. Fortunately my childhood schooling had accustomed me to being picked first in any alphabetical ordering. The mayor announced that each attendee between A and Z would also receive a certificate from the desk outside. Mine still hangs in my office.

Without question, pride of place at that meeting went to Ian McLean's talk, lavishly illustrated with pictures taken using the new infrared camera that had been operating for several months in one of the white domes atop Mauna Kea, the craggy volcano that looms over the town. Regular *Yearbook* readers will recall Ian's article in the 1989 issue. But nobody who was not intimately involved can begin to understand the excitement we all felt as slide after slide of Ian's appeared on the big screen, each transforming another of our dreams into real data.

Ian had chosen indium antimonide, the alternative detector material. But the Rockwell Science Center had made great progress in the intervening six months, and my preference remained for the newer option. By the end of the conference I was ready to ask for a contract with Rockwell.

Rejection, 1988

Things moved unbelievably slowly. The cost of one of the detectors was high enough that I had first to convince my superiors that I was making the right choice. Then my contact at Rockwell had to convince his superiors that it was OK to deal with someone in Australia. Finally, he had to seek an export licence for official permission to sell a detector abroad. It was therefore not until half

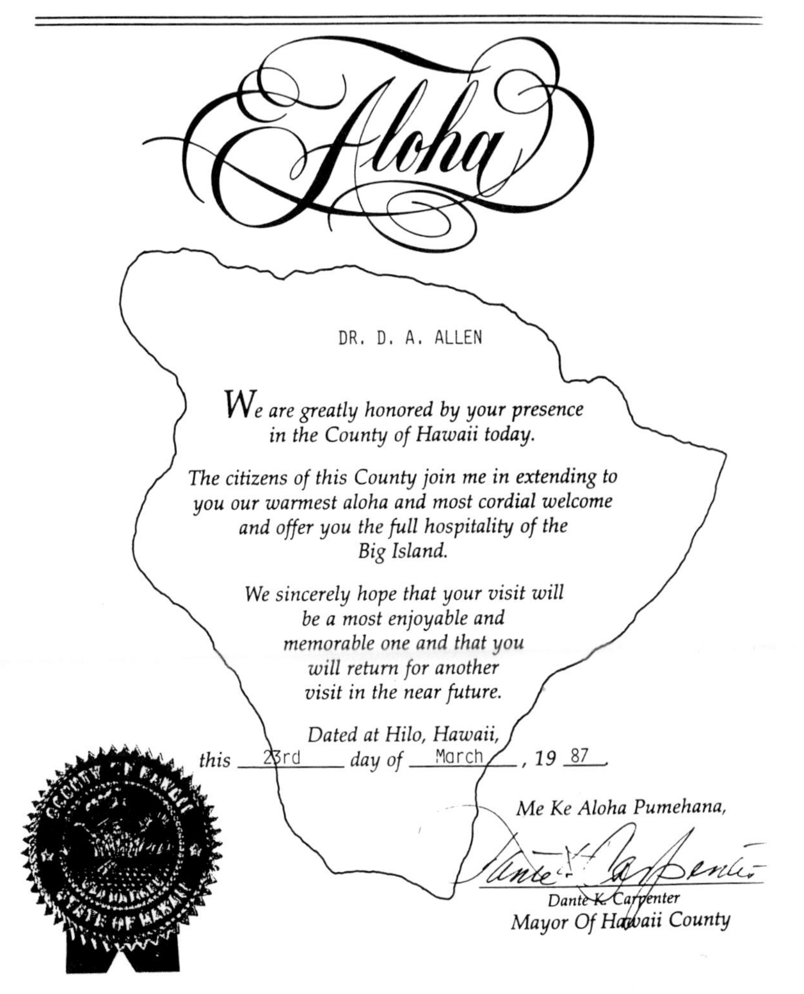

COUNTY OF HAWAII

Aloha

DR. D. A. ALLEN

We are greatly honored by your presence
in the County of Hawaii today.

The citizens of this County join me in extending to
you our warmest aloha and most cordial welcome
and offer you the full hospitality of the
Big Island.

We sincerely hope that your visit will
be a most enjoyable and
memorable one and that you
will return for another
visit in the near future.

Dated at Hilo, Hawaii,
this 23rd day of March, 19 87.

Me Ke Aloha Pumehana,

Dante K. Carpenter
Mayor Of Hawaii County

Figure 1. The certificate presented to the author at the 1987 conference by the Mayor of Hawaii, Dante K. Carpenter. Aloha *is a Hawaiian greeting literally meaning 'love' but generally used as 'hello'.*

way through 1988 that the stunning blow was dealt to our ambitions.

Technology transfer was the term used. But call it what you will, the basic issue was caution over letting out of the USA any remotely sophisticated equipment. Someone, somewhere, thought these detectors might have military significance, and vetoed the sale. We knew, and the manufacturers knew, that that was a fallacy, but the decision was out of our hands. The licence was denied, and the decision was deemed by knowledgeable people irreversible.

Many offered advice, and had I taken it all we would now be operating an indium antimonide camera like Ian's, or perhaps some other device. But there's a stubborn streak in me. I knew that mercadtel suited our needs better, and by golly I was going to get it. I therefore sought a fairy godmother.

One appeared in the unlikely guise of Bill Lane, millionaire publisher of the *Sunset* books. No, this wasn't a case of buying one's way out of difficulties. Bill Lane was serving as US ambassador to Australia. He is a man with a keen interest in science. During the time he was in Australia he donated tens of thousands of dollars of his own money to support local scientific endeavours, exhibitions and the like. Bill saw our case as worth a fight. He and his team swung into action, contacted people in very high places, and after twelve months work had performed the impossible task of having the rejection reversed. Whenever I see a *Sunset* book now I think lovely thoughts of Bill Lane, a man whom I never had the opportunity to meet personally before his term was up and he returned to California. I hope I get the chance to thank him personally one day.

I should add that even Bill Lane might have failed had Australia not become a signatory to a scheme known as COCOM. I won't bore you with the details of COCOM, save to say that it was intended as a means of easing technology transfer out of the USA. I mention it because there is a feeling in some quarters that COCOM introduced a whole lot more red tape with no benefits. For us the reverse was true.

Acquisition, 1990

Still there was a dragging of heels in those halls of power in Washington, or wherever. Though the export licence was promised in mid-1989, it didn't eventuate until January 1990. Only then could Rockwell proceed to make our detector, and that required longer than expected. We finally took delivery in September 1990, three and a half years after I had first approached Rockwell's key man, Kadri Vural.

There was a bonus, however. Just as the Rockwell devices had greatly improved during the six months between my Californian and Hawaiian visits, so they continued to improve. The detector we acquired in 1990 was a great deal better than we could have bought a year or two earlier. Moreover, once we were certain the device was coming we had time to design and build a sophisticated instrument to house it in. Not for us the simple camera that can fire off a few snapshots of the infrared sky. We aimed to produce a first-class instrument, and I am conceited enough to suggest that we succeeded.

I could bore you again, this time with a score of names of key people who made our instrument what it is. There are too many. Though I was the project scientist, and therefore received all the accolades, the truth is that a great deal of work was needed by a top-class group of engineers and technicians. I marvel to recall that only three months after we received the detector we had worked out how to get the best out of it, had installed it in the instrument, and had taken the whole massive construction to the telescope to try out. IRIS, our infrared camera and spectrograph, saw first light in December 1990 on an evening I will long remember for its adrenaline high. By March we were giving the first outside users access.

IRIS

Nor, if I want to keep my readers, should I indulge in a lengthy technical description of the instrument we built. On the other hand, I can't resist a few details. The name, for instance. Quite early on I was asked to give the instrument a name. It has become traditional of late to use clever acronyms for astronomical instruments. Thus we find such as HIFI, FLAIR and 2D-FRUTTI. One group concocted the acronym PEPSI for their new instrument, only to learn that they would be sued by the manufacturers of a certain beverage if they ever used the name in print.

Walking home one evening I pondered names beginning with IR, for infrared, and suddenly found IRIS in my mind. I became aware of some of the sick puns one could make during the development phase – taking root, in the bud, blossoming, etc. These didn't put me off, so IRIS it became. People ask me what the second I and the S stand for, but as I conceived the name they didn't stand for anything. Since that isn't an acceptable answer nowadays, let's agree on *Imaging* and *Spectroscopy*.

Imaging is, of course, what one expects an infrared camera to do. We designed IRIS to be versatile, to take pictures at different scales. The coarse, wide-angle mode would give a feel for an object; the detailed telephoto view would show detail. This meant a selection of lenses and mirrors, with machinery to move them in and out of place by remote control. All observing is done these days from a control room, and it would be inefficient to have to go out to the telescope every time the observer wanted to change something in IRIS.

But a complication of infrared detectors is that they, and all the pieces around them, have to be kept very cold. The IRIS detector works best at about −190 degrees C (−310F). Now, observatories can be pretty chilly places at night, but they don't get *that* cold. So the detector, optics and mechanisms had to be built inside a huge cylindrical box with all the paraphernalia to generate such low temperatures. We use a type of refrigerator that differs from the one in your kitchen mostly in that it pumps helium gas around instead of freon. There's no damage to the ozone hole if IRIS's refrigerator leaks.

The pump is a separate box the size of a small table and much heavier than me, and it must stay upright, and be connected to IRIS by short, flexible pipes. Most astronomers find it inconvenient to use a telescope that can't look anywhere but straight up, so we had to find a way to tip the telescope without tilting the pump. Our solution was to hang the pump from a wire loop at the end of a long arm bolted to the telescope. The pump is always right way up wherever the telescope points. You have to walk under this huge, hanging pump if you need to visit IRIS when she's on the telescope, and believe me it has taken me a while to do so without hesitating, looking up, and sprinting.

The pump sends a spurt of helium gas every second, and it makes a sort of whooshing noise in the pipes. When IRIS is in use her heartbeat echoes through the dome in what I find to be a re-assuring way. Others refer to her heavy breathing.

The different image scales aren't the only bits of optics inside IRIS. To the Romans Iris was the goddess of the rainbow, so we felt we needed a way of splitting up the infrared radiation into its wavelengths. In other words, of taking spectra. We devised a novel way of doing this, and as a result have a much more potent instrument. You can take a picture of a chunk of sky; decide there's an unexpected feature in it; put in the telephoto option for a better look; reckon the feature is worth more exploration; then swing in the spectroscopic option and thus learn something of the physics or the chemistry of it, just as one does in optical astronomy. The ability to do spectroscopy elevates IRIS from a simple camera to a sophisticated instrument.

Discoveries

Enough of the instrument. What has it shown us? Well, needless to say, a good many things. A few of the pictures IRIS has produced are scattered through this narrative, and their captions will give you a taste of how infrared astronomy reveals different features from optical. But let me return to the four thoughts that crowded my mind as I started to write this piece.

Centaurus A was our first real discovery. It is a big galaxy, and therefore a good object to test how well IRIS handles an object bigger than she can view in a single image. We took four images that we could overlap to make one big picture. The optical photograph of Cen A (as the galaxy is generally known) is well known for the braided lane of dust that crosses the centre of the galaxy. The dust is mixed with gas and lies in a warped pancake-shaped layer that meanders through the galaxy and is viewed almost edge-on.

I seem to recall that it was Michael Burton who suggested we take the pictures. Vikki Meadows and I watched as he fired them off. But when the images came up on the screen we were stunned to find that far from being a dark band, the dust lane had in parts become bright! We took the data back to think about them, and quickly realized that we needed better images. The picture shown here (Figure 2) is one of a set that Vikki and I made a few months later. As you can see, we took five overlapping images and pieced them together.

Why has the dark dust turned bright? The several possible answers to this question worried us for a while, but with some careful analysis of the images at several wavelengths we have been able to show that in fact many stars have formed within the dust lane. These are hidden from view at optical wavelengths by the dust itself, but dust becomes more transparent in the infrared, so IRIS records some of the underlying light.

It is generally believed that the girdle of dust around Cen A is the

Figure 2. The galaxy Centaurus A is a huge, almost spherical ball of stars threaded by a dust-filled wafer of gas from a spiral galaxy that it once consumed. The dust becomes partly transparent at infrared wavelengths, and reveals bright patches, on either side of the galaxy, due to stars mingling with the gas.

relic of a spiral galaxy that was swallowed by the big elliptical some hundreds of millions of years ago. The gas and dust of the spiral's arms would settle into the pancake we now see. In so doing the gas would be compressed, and that is usually a good reason for stars to form. The stars that IRIS revealed must date from after the merger of the two galaxies.

The images of Cen A show a great deal more than this, but if I detailed it all there would be no room for my other scientific snippets.

Getting out of order let me deal with the mass-loss star now, for there is a moral to the tale. No sooner had IRIS started to take pictures than the media wanted stories. Peter Pockley, one of Australia's foremost science writers, telephoned and asked if his newspaper could publish a picture taken that week. I thought quickly and remembered an object that has been given the unfortunate name of the 'rotten egg' nebula because of the pongy gas that radio astronomers have found in it. Vikki, who has a more romantic view, dubbed it 'the kiss', for it looks rather like two elongated faces almost touching. This appearance arises because a very old star is shedding its outer layers along two opposite directions to make elongated nebulæ. The gas escapes in those opposite channels because a very gooey disk of dust and gas surrounds the star, preventing gas escaping past it and actually hiding the star completely from view. That is the slender gap between the osculating lips. Nobody is really sure of how such a huge disk could be produced, and one idea has it that the disk was there from the outset – a planetary system that never managed to form.

I had time for only a quick analysis of the images to meet Peter's deadline. His editor never used the picture anyway, so the data were set aside until recently when I worked through them again. To my amazement I found the elongated nebula surrounded by a very faint, but perfectly circular, outer nebula. The only way a circular nebula could have formed around it is if there had once been a spherical outflow, unhindered in any direction.

Now, there is no known example of an object that has changed from a spherical outflow to the strongly directional form. Here was proof that the constraining disk had formed recently in the star's history. Now, at last, I could put real limits on how such a disk might have formed. I was elated.

All good discoveries need to be confirmed, of course. I re-observed it only a week or two before I started to write this. The circular nebula appeared more clearly than before. More, it was slightly bigger. Now, I know that it couldn't have grown that quickly: the gas would have needed to travel faster than light.

Suddenly the truth dawned. This is no real nebula at all – it is an unexpected reflection inside IRIS.

Well, that's my current interpretation. I shall test this idea next time I have a chance by observing a very bright star. But unless I add a postscript to this story, you should assume that this was another of those examples of the vagaries of science. Another great discovery to be dropped into the circular file beside my desk.

That is not the case of the spectroscopy, however. Since its eruption in 1987 a team led by Peter Meikle of Imperial College has been following the changing spectrum of supernova 1987A. Jason Spyromilio and I were two devoted members of this team, and together with some of Peter's students we have made some really exciting measurements. The infrared wavelengths accessible to IRIS have proved the most powerful in understanding supernovæ.

But as the supernova faded, the measurements became harder and harder. Though we were anxious to see how the supernova evolved, the existing infrared spectrograph just didn't have the sensitivity to do the job. We had to abandon the work.

So it was with considerable anticipation that we trained IRIS on supernova 1987A, now a faint smudge of light much dimmer than the star had appeared before it erupted. It is so faint, in fact, that its radiation is swamped by that from the sky, so that we couldn't tell what we were learning about it as we observed. We spent about three hours taking data, then I sat down to work through the analysis while Jason turned to a different supernova. Within half an hour I had our spectrum analysed, and I plotted it on the screen. We both stopped in our tracks. The quality of the wavy yellow line was as good as we had been getting two years before with our old instrument. We were back in business.

We sent an immediate congratulatory electronic mail message to Peter, who had been unable to come to the telescope. We soon followed this up by sending the data. I'd like to be able to say that Peter was knocked off his chair by the result, but one of the troubles of electronic-mail communication is that you never find out whether something like that happens. Still, he ought to have been.

IRIS has made it possible to get valuable data on other supernovæ, too. Conveniently, supernova 1991T erupted just after IRIS was brought into service. Our initial spectrum of it was taken before the supernova had reached maximum light, and was the first such infrared spectrum of a supernova of this type. The graph shows a plot of the first two spectra. Don't worry if graphs like this leave you cold – we haven't fathomed this one out either. Suffice to say that the April and June observations are strikingly different, and that's enough to set astronomers gossiping. It may be a few years before the change is understood, but the data are just beautiful.

Beauty is a term one associates also with my final example: the planet Venus. And it is where Dave Crisp enters the story again.

Some years ago I had discovered a way to study the deep atmosphere of Venus by taking infrared images of the dark side when the planet has a phase like a crescent moon. The surface and lower atmosphere are so hot that the heat they emit penetrates the tenuous clouds and can easily be detected by a small telescope equipped with infrared gear. Where sunlight illuminates the upper clouds of Venus the glare reflected back completely swamps the deeper emissions, so the observations can be made only when Venus reaches a suitable phase, for a few weeks every nineteen months. Previous to this discovery it had been thought that the atmosphere of Venus was amenable to study only by visiting spacecraft.

We were anxious to see what IRIS could contribute. Dave came to Australia to observe the planet with me a few months after IRIS had been brought into commission. By using the spectroscopic mode we could map the distribution of gases such as water vapour only a little way above the surface of the planet. This is important to understanding how the atmosphere functions. Venus is extremely dry: we find only about 40 parts per million of water. It never rains there, and there is no surface water, so that the atmosphere is simpler than the Earth's. If we can understand the Venus atmosphere, then we will have a better basis for understanding our own, more complex atmosphere.

The observations could be made in daylight, but we were interested also in exploring some other wavelengths where I had recently found that the dark side also gave out infrared radiation. To do so we had to await sunset. The trouble was that there was only about half an hour to make the observations between sunset and the time the planet sank too low for the telescope to track. This is just the time of day when clouds often peel off the surrounding mountains before vanishing into a clear night. Half of the available time was lost as clouds drifted by, and our first glimpse of the planet came with just seventeen minutes to go.

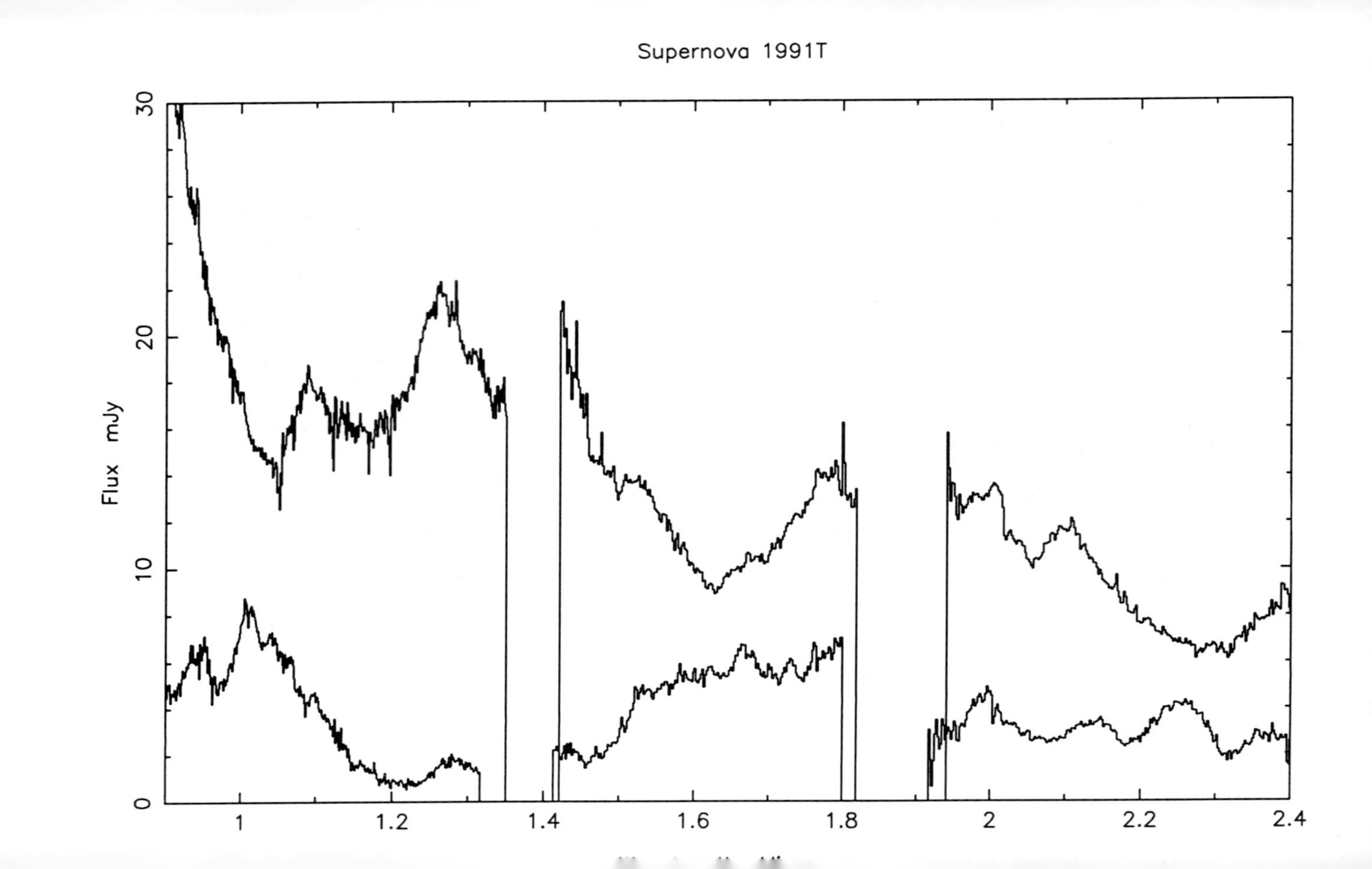

Supernova 1991T
Flux mJy
0
10
20
30
1
1.2
1.4
1.6
1.8
2
2.2
2.4

Immediately we both noticed something strange. There was a bright patch near one edge of the planet at the exact wavelength of a feature due to oxygen molecules. Oxygen is known at the top of the Venus atmosphere: it forms when sunlight decomposes carbon dioxide. It is a weak glow very similar to the pale glimmer of the Earth's night sky. The bright spot we saw certainly wasn't how it was supposed to look.

In seventeen minutes we could just about get the map we clearly needed to show where the glowing oxygen really lay. Here was a discovery just waiting to be recorded. I typed 'run' into the control computer, and IRIS whirred into action.

Immediately the screen flashed up an array of warning messages. Somewhere, deep in the interface between electronics and computer, something had gone badly wrong. IRIS wouldn't work.

There are various courses of action under these circumstances. You can leap up and down fuming or sobbing according to your preference, and hurl abuse at all computers and instruments ever built. You can even aim the occasional kick at less sensitive bits of the equipment. This course, however, never gets you your data. You can go off and get drunk, but that's just as ineffective. More constructively, you can try to coax IRIS with a few knee-jerk bits of software; each takes a couple of minutes to run, but none is guaranteed to work. Or you can patiently switch everything off, count to ten, switch it on again and start afresh. That takes about ten minutes. There is a risk that even this won't cure the symptom, that IRIS will wake up still paralysed.

I switched everything off.

Try, if you can, to put yourself in our position. Slowly, one by one, the bits of software announce on the screen that they are being fired up. Quickly, so quickly the minutes are counting down on another screen that shows how much longer the telescope can continue to track Venus before it runs its limit. You sit, helpless, unable to do a thing, willing the Earth to stop turning so that Venus would cease its relentless descent into the west.

With less than eight minutes to go IRIS woke up. But not the way we had set up before, for a number of mechanisms settle into a wake-up state. There followed a spell of frantic typing, trying

---

Figure 3. These wiggly graphs show how the spectrum of supernova 1991T changed between April and June 1991. The earlier spectrum is the higher plot. The gaps in the coverage are regions where the Earth's atmosphere does not allow radiation to pass.

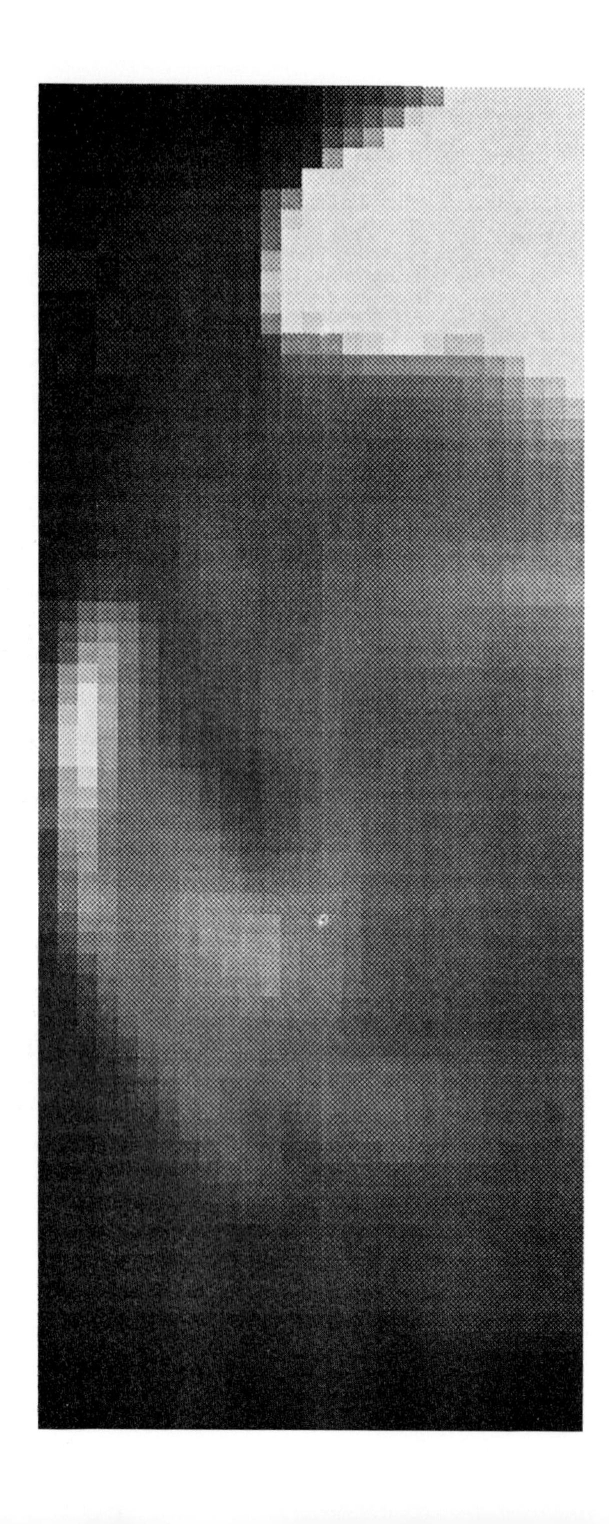

desperately not to make a mistake, striving hard to remember all the changes that we had made. At last, believing everything to be right, I typed 'run' again.

IRIS ran. The data came in. We had to cut the data flow short to get any map at all, and still the telescope reached its limit, and stopped following Venus, before we had finished. But we had enough to show our newest discovery (Figure 4).

We observed again on several further occasions. One of them was even more phrenetic. Our discovery was confirmed, and to add

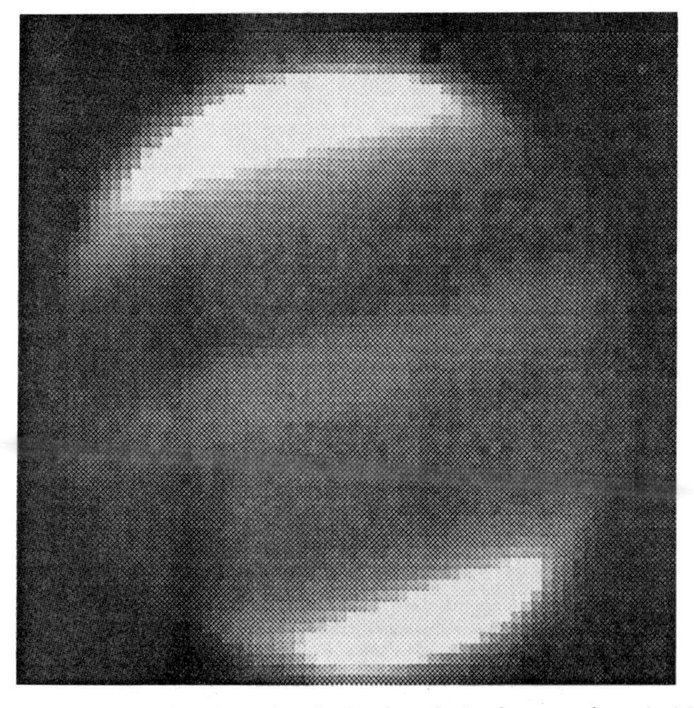

Figure 5. At some infrared wavelengths the planet Jupiter becomes almost invisible because methane gas in its atmosphere absorbs sunlight. In this image the faint equatorial belt and the bright poles are deficient in methane.

Figure 4 (opposite). Oxygen auroræ show on the dark side of the planet Venus. The sunlit crescent is grossly overexposed at the top of the image.

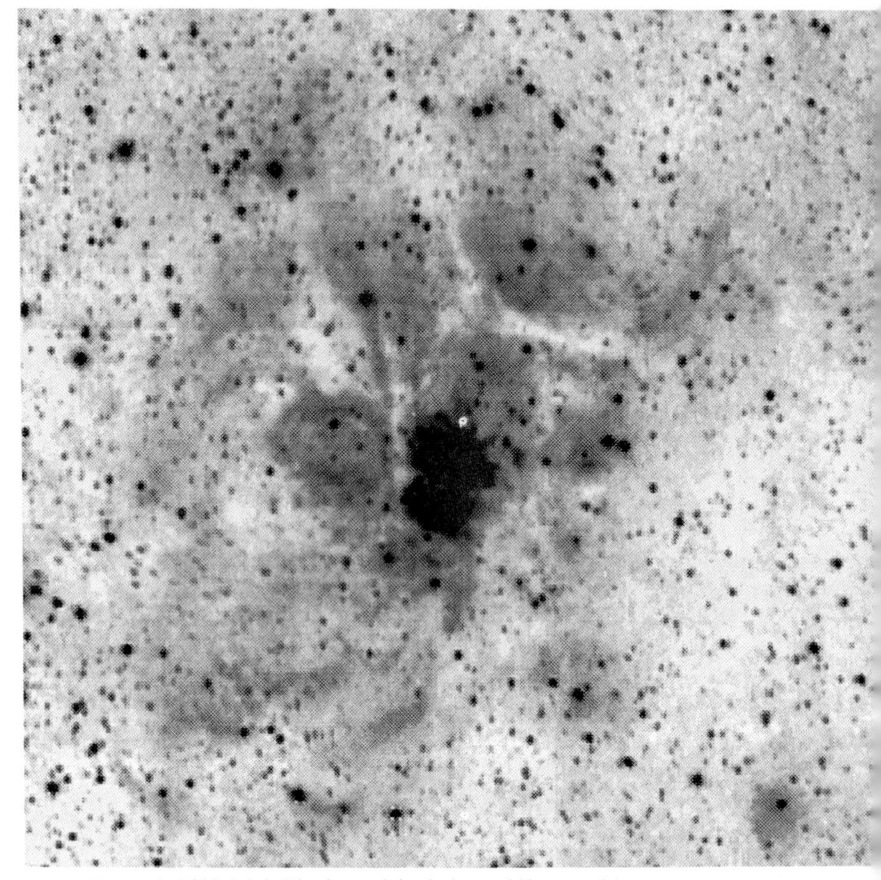

Figure 6. G333.6-0.2. This beautiful nebula would be one of the most prominent in our skies but is completely hidden from view by the dust clouds that throng our Galaxy. It lies in the southern constellation Norma. Radio astronomers discovered it, and gave it its uninspired name. (This is a negative picture.)

interest the oxygen emission changed from night to night. More often than not there was no really bright spot. If that had been the case on the first night we would have abandoned trying to get any data in those last eight minutes, and the discovery might not have been made.

What *did* we discover? The simplest description is to claim that we found giant auroræ on Venus. Here on Earth we see auroræ near the magnetic poles. They are caused by oxygen and nitrogen gas interacting with charged particles given off by the sun. But they occur, spectacularly sometimes, only because the magnetic field funnels the particles into dense accumulations, and the auroræ form only where those accumulations are made.

On Venus the bright oxygen feature doesn't lie near the magnetic poles, but in the tropics close to the middle of the Venus night. In fact, Venus has essentially no magnetic field at all, so that true auroræ are impossible. What causes the strong oxygen glow? Would that I could tell you, but at present we are not really sure. As Dave Crisp tells me, we have opened a can of worms. Our best guess is that we have uncovered a vast circulation pattern of oxygen at the top of the Venus atmosphere, and the bright features show where oxygen is falling back down into the upper air of the planet.

Dave takes me to task for referring to this as an aurora, yet the variable, streaky patterns we have found resemble auroræ on Earth. An observer on Mars studying the Earth would record virtually the same appearance during an auroral display as we see when Venus is active, though nearer to the Earth's poles. So, if your friends ask you what this new-fangled infrared camera has discovered, tell them it's found giant auroræ on Venus.

Symbiotic Stars

JOHN ISLES

Biologists use the term *symbiosis* ('living together' in Greek) to refer to a partnership between two organisms of different species that benefits both organisms. An example of symbiosis is given by the lichens, the plants that form crusts and tufts on stones, trees, and soil; each type of lichen actually consists of a fungus and an alga living symbiotically. Another example of symbiosis is given by dogs and (some) humans!

Just over half a century ago, the American astronomer Paul W. Merrill introduced the term *symbiotic stars* to describe stars of two very different kinds that seemed to occur together, as if they needed one another. He was observing pairs of stars that appeared as single stars in his telescope, but whose spectra revealed the presence of two components: a cool, red giant or supergiant star, and a hot object, which has generally been interpreted as a companion star, revealing its presence indirectly by raising part of the red giant's extended atmosphere to a high temperature. The low-density hydrogen-rich gas that is heated in this way emits radiation at certain fixed wavelengths, so that the spectrum shows bright emission features such as the red line of hydrogen-alpha. The hot object may also be detected directly, if its own brightness is high enough, by contributing extra light at the blue end of the continuous part of the spectrum.

Single stars such as our Sun can behave in ways that are difficult enough for astronomers to understand. Close binary stars such as the symbiotics are even more challenging. Consider how complicated their variations in brightness can be. First, the red component is very often a pulsating variable star – either a Mira star or a semi-regular variable. Mira stars, named after the prototype Mira (Omicron) Ceti, are also called long-period variables. They are red giants that pulsate with periods of typically about a year, and a range in visual brightness of two and a half magnitudes or more. Semi-regular variables are similar, but have a less clear period, or have two or three periods present at the same time. Their ranges of brightness variation are generally much smaller than those of Miras.

Second, the symbiotic system does not radiate light equally in all directions, because one side of the red giant is heated by the hot star. As the stars orbit about their common centre of mass, with periods of several months up to some years, their combined light as seen from the Earth varies in the same period – unless we happen to be viewing the symbiotic pair pole-on. Also, tides raised on the red star by its companion cause it to be egg-shaped, with changing visible surface area as the stars turn in space. Changes of this type are called *ellipsoidal* variations; there are two maxima and two minima in each orbital period, when we see the red giant side-on and end-on.

If the orbit happens to be orientated in such a way that we are near its plane, eclipses may also be seen as one star passes in front of the other. When the hot star passes across the face of its much larger companion, there may not be much change in the combined light. (There are possible exceptions to this rule, as we shall see later in the case of R Aquarii.) A deep minimum is sometimes seen, however, when the hot star is eclipsed by the red giant.

The hot component in symbiotic systems is thought to be feeding off its partner, either collecting material from a nebula that surrounds the red star, or syphoning it off the red star's surface in a stream that feeds what is called an *accretion disk*, a whirlpool of gas spiralling in towards the surface of the hot star. Owing to the great light output of the red giant or supergiant component, the hot companion and any accretion disk that may be present cannot often be readily detected in the visual spectrum, but measurements at shorter wavelengths show an excess of ultraviolet light attributable to the hot star and its disk.

In at least some symbiotic stars, the hot star seems to be a white dwarf. Outbursts are seen that resemble those of novæ and dwarf novæ, which are interacting binaries comprising a cool star and a hot one. The cool star in novæ and dwarf novæ is an ordinary main sequence star, similar to our Sun but somewhat cooler; it is not a giant nor supergiant as in the symbiotics. The hot star in novæ and dwarf novæ is a white dwarf: a small, dense star that has exhausted its hydrogen fuel and is in the final stage of its evolution. The rare outbursts of novæ are due to an explosive thermonuclear reaction, the sudden 'burning' of hydrogen to helium in the material the white dwarf has collected from its partner. The dwarf-nova type of outburst is more common. This is the pulsed release, in the form of light and heat, of gravitational energy from material in an accretion disk that is falling towards the white dwarf. At different times, the same star can show both nova and dwarf-nova outbursts.

In symbiotic stars, similar outbursts apparently occur but usually they do not progress in the same way as those of novæ or dwarf novæ. Several symbiotics have undergone very slow, nova-like outbursts that lasted years or decades, whereas ordinary novæ fade back to minimum in a matter of months. Ordinary novæ are thought to recur on a timescale of many centuries, but two symbiotics (T Coronæ Borealis and RS Ophiuchi) are recurrent novæ, with respectively two and five recorded outbursts, at intervals ranging from nine to eighty years. Many symbiotic stars have also shown irregular or quasi-periodic eruptive activity. In symbiotic pairs that are eclipsing binaries, the eclipse can be spectacularly deep if the hot star or its accretion disk is in outburst when it passes behind the red giant.

Although such complicated processes as we see in symbiotic stars are hard to unravel, astronomers study them in their quest for understanding of everything in the Universe. The study of all kinds of stars tests our knowledge of how stars work and increases our understanding of the operation of the laws of physics. If symbiotic stars are different from other interacting binaries, this must be because they present unique physical conditions. Moreover, we can expect to learn much more from studying a star that changes, especially in a complicated way, than from one that does not; and even if we cannot interpret all the observations now, it is still important to record these stars' activity for the benefit of future generations of astronomers. There is great satisfaction to be gained from the knowledge that one's own observations may be useful to astronomers a century from now.

The evolution of symbiotic stars

Our present understanding allows us to describe in outline how a symbiotic star may form, and what may be its destiny. Future research will undoubtedly change many of the details of the story, and we should bear in mind that the symbiotic class may include some very different kinds of objects from those we shall describe.

The pair of stars that will eventually become a symbiotic system must have begun life as a normal double star, containing two main-sequence components, each much like our Sun but probably with somewhat greater masses, perhaps about three times the Sun's mass. At this stage, each star obtains its energy by nuclear fusion in its core, converting hydrogen to helium.

When the hydrogen in the core is exhausted, each star begins hydrogen 'burning' in a shell surrounding the core. This happens first in the more massive of the pair, since massive stars consume their resources more quickly. The outer layers of this star, which we shall call the primary star, expand and its surface cools, so that it becomes a red giant and probably passes through a phase of pulsation as a Mira star. If it were a single star, it might eject its outer layers as a planetary nebula. Instead, its companion star, the secondary, picks up this lost mass.

While the primary star's core, which is all the matter it now has left, settles down to become a white-dwarf star, its companion, which by now may have become the more massive of the two (but which we shall continue to call the secondary), begins its own evolution to the red-giant stage, and in its turn also becomes a Mira star. In the well-known Hertzsprung-Russell diagram that shows the relation between the luminosities and temperatures of stars, the secondary is located on the *asymptotic giant branch*. This is a phase in which the star expands and cools while maintaining nearly constant average luminosity. As it expands, the period of oscillation increases, and the star's mass-loss rate increases dramatically.

The lost mass forms a nebula around the secondary, part of which is excited to emission by radiation from the white-dwarf primary. When the nebula has become dense enough for bright lines to be visible in the combined spectrum of the binary star, astronomers are able to recognize the binary as a symbiotic system.

At a later stage, the red giant's mass loss may accelerate further to become a 'superwind', leading perhaps to the formation of a planetary nebula. When this nebula has dispersed, we are left with an inert pair of white-dwarf stars. Over many millions of years, they will lose their orbital energy by emitting gravitational radiation, and will gradually spiral towards one another, ultimately merging in one type of supernova explosion.

The story we have told assumes that the red giant's companion in a symbiotic system is a white-dwarf star. Analogy with the novæ and dwarf novæ, which certainly have white-dwarf components, strongly suggests this is the case, but it is difficult to be certain from observations of symbiotic stars. In an alternative model, the hot star is an accreting main-sequence star. If this is correct, then symbiotic binaries are much younger systems, and the cool star is the more massive component and the first to evolve to the red-giant stage. In this model, the system's final fate as a supernova is unchanged. Perhaps symbiotic stars of both types exist, some with white-dwarf companions and some with main-sequence companions; we do not know yet. At all events, it seems that, far from being of mutual benefit to both stars, the symbiotic relationship leads in the end to mutual destruction.

Some bright symbiotic stars

Table 1 lists a selection of bright symbiotic stars that vary in brightness and are easily visible in amateur telescopes. There are

TABLE 1

Some Bright Symbiotic Stars

Star	*R.A.* (2000.0)		*Declination*		*Range*	*Period days*
	h	*m*	*deg.*	*min.*		
Z Andromedæ	23	33.7	+48	49	8.0–12.4*p*	632
EG Andromedæ	00	44.6	+40	41	7.1–7.8*v*	470
R Aquarii	23	43.8	−15	17	5.8–12.4*v*	387
UV Aurigæ	05	21.8	+32	31	7.4–10.6*v*	394
TX Canum Venaticorum	12	44.7	+36	46	9.2–11.8*p*	—
T Coronæ Borealis	15	59.5	+25	55	2.0–10.8*v*	29000
BI Crucis	12	23.4	−62	38	11.0–14.0*p*	—
BF Cygni	19	23.9	+29	40	9.3–13.4*p*	755
CH Cygni	19	24.6	+50	14	5.6–9.2*v*	97
CI Cygni	19	50.2	+35	41	9.1–11.5*v*	855
V1016 Cygni	19	57.1	+39	50	10.1–17.5*B*	—
AG Draconis	16	01.7	+66	48	8.9–11.8*p*	554
NQ Geminorum	07	31.9	+24	30	7.4–8.0*v*	70?
RW Hydræ	13	34.3	−25	23	8–9*v*	370
SS Leporis	06	05.0	−16	29	4.8–5.1*v*	—
BX Monocerotis	07	25.4	−03	36	9.5–13.4*p*	1374?
RS Ophiuchi	17	50.2	−06	42	4.3–12.5*v*	—
AR Pavonis	18	20.5	−66	05	7.4–13.6*B*	606
AG Pegasi	21	51.0	+12	38	6.0–9.4*v*	800
AX Persei	01	36.3	+54	16	8–13*v*	682
RX Puppis	08	14.2	−41	42	9.0–14.1*B*	580
HM Sagittæ	19	42.0	+16	45	10–17*v*	550
FN Sagittarii	18	53.9	−19	00	9–13.9*p*	—
V1017 Sagittarii	18	32.1	−29	24	6.2–14.9*B*	—
FR Scuti	18	22.7	−12	44	10–12*v*	—
RT Serpentis	17	39.9	−11	57	10.6–17.0*p*	—
FG Serpentis	18	15.1	−00	16	9–13*v*	—
RR Telescopii	20	04.3	−55	44	6.5–16.5*p*	395
KX Trianguli Australi	16	44.6	−62	37	10.1–13.6*B*	—
PU Vulpeculæ	20	21.2	+21	34	8.7–16.6*p*	—

many more symbiotic stars that are not yet known to vary, but this is probably simply because the observations to check for changes in their magnitudes have not yet been made. The first column lists each star's name in the *General Catalogue of Variable Stars* (GCVS), compiled in Moscow. The positions are given in terms of right ascension and declination for epoch 2000.0. Visual (v) magnitude ranges are given where possible, but some ranges are in blue light (B) or derived from blue-sensitive photographic plates (p). These may be somewhat fainter than the corresponding visual ranges.

Except when they are at faint minima, all the stars listed should be identifiable using a telescope of 100-mm aperture and a good star atlas such as *Uranometria 2000.0*. Several of these stars are usually within range of binoculars, and can be found with the help of *Sky Atlas 2000.0*. If there is a known period of variations, this is listed in the table, but the periodic changes do not always account for much of the light variation, as more than one period may be present, and there are usually many irregularities. The period quoted may be that of pulsation, that of orbital motion, or the average interval between outbursts. Some symbiotic stars of special interest are described in the following notes.

EG ANDROMEDAE lies in the direction of the Andromeda Galaxy (M31), but it is actually a member of our own Galaxy, only some 500 parsecs distant. (One parsec is 3.26 light-years, or about 31 million million kilometres.) Its visual magnitude is normally near 7.5. Slight variations were first reported by the Czech astronomer T. Jarzebowski in 1964, but the star's unusual M-type emission spectrum had been noted as early as 1950. Oscillations ranging between 0.1 and 0.3 magnitude have been reported with a period of 40 or 80 days; these may be due to pulsations in the red giant. A 470-day period, with range of 0.2 magnitude, has also been found from an analysis of visual observations by the American Association of Variable Star Observers; this variation seems to be linked with the orbital period. There is independent spectroscopic evidence for eclipses of the hot component by the giant star every 470 days. The variations of EG Andromedæ are rather small for visual study, and it is a good candidate for photoelectric photometry. An attraction of this star is that it offers alert observers a good chance of discovering any supernova that may occur in M31, which lies in the background about 670,000 parsecs beyond EG Andromedæ.

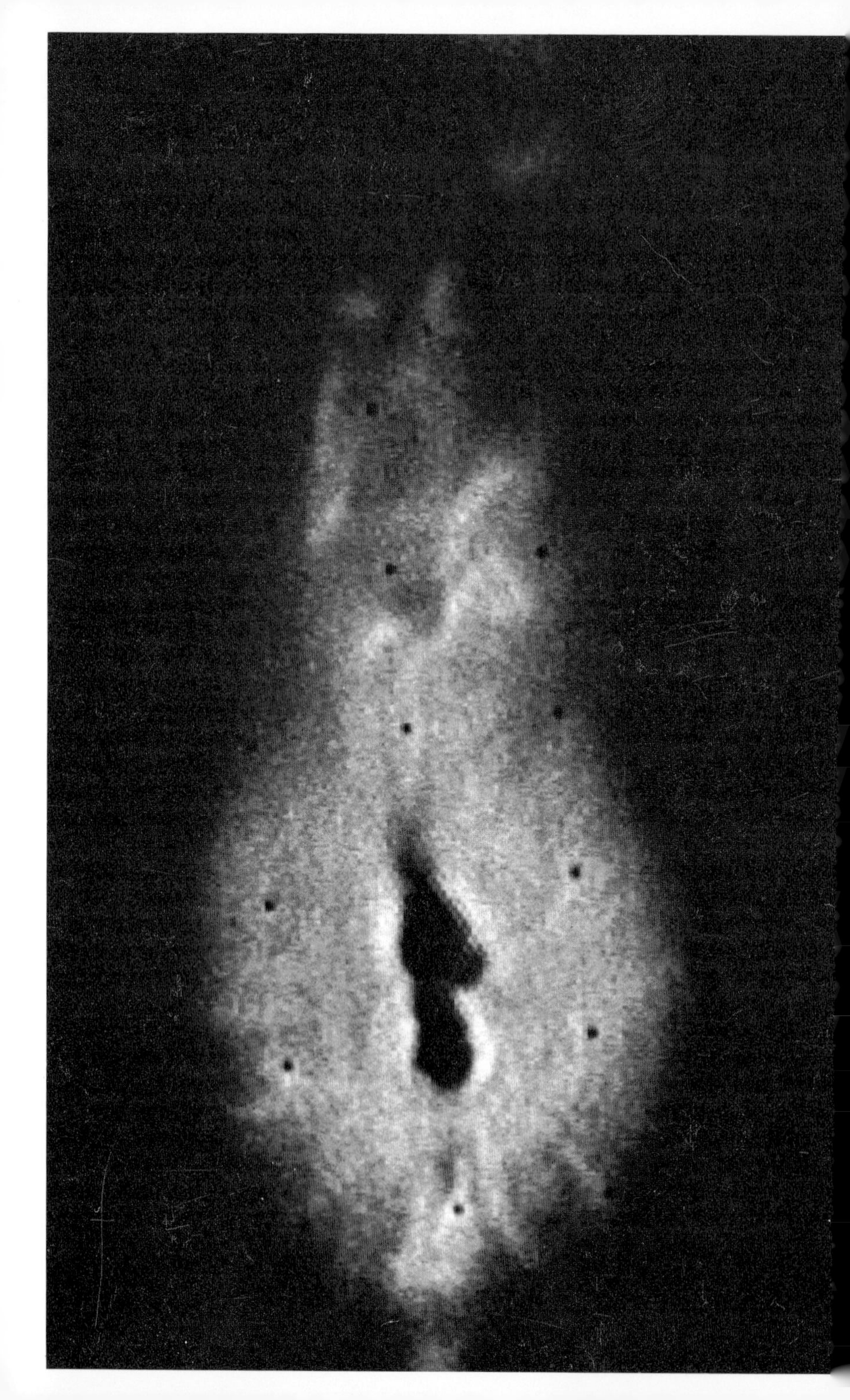

R AQUARII is a well-known Mira star (Figure 1). Its wide variations in an eleven-month cycle were noted in 1811 by K. L. Harding in Germany. Between 1928 and 1937 its variation range was smaller than normal, and in 1932–34 it hardly varied at all. In the late 1970s there was a second episode of unusual behaviour, when the maxima and minima became fainter than normal. These two episodes, separated by about 44 years, have been interpreted as eclipses of the Mira component by a dust cloud surrounding the hot companion, which was supposedly undergoing a long, bright outburst on the earlier occasion. An orbital period of 44 years has been suggested. Few observations were made around the times of earlier possible eclipses in 1845 and 1889, so we may have to wait until the next one in 2022 to test this theory. R Aquarii has a combination spectrum characteristic of symbiotic stars, and it is an X-ray and radio source, with jets that may have been ejected from an overloaded accretion disk in the 1970s. It is surrounded by a faint, expanding nebula that may be the remnant of a nova outburst in 1073 recorded by Korean astronomers.

At declination −15 degrees, R Aquarii is best observable from mid-northern latitudes when it is near the meridian. Its next maximum is predicted for mid-January 1993, when it should be easily visible with binoculars in the south-west as darkness falls.

T CORONÆ BOREALIS is the famous recurrent nova of 1866, when it was discovered by the British astronomer John Birmingham, and 1946, when two British amateurs independently found it again in outburst. The GCVS lists its period as 29,000 days, which is the interval between the known outbursts, but there was another possible sighting by John Herschel in 1842, so the true period may be shorter and the next outburst could come at any time. T Coronæ

Figure 1. This is the first picture taken by the Hubble Space Telescope of the symbiotic variable R-Aquarii. The two dark knots at the centre of the image probably contain the binary star system itself, which consists of a red giant and white dwarf star. The knots are dark due to saturation effects produced by the FOC detector when it observes very bright objects.

The filamentary features emanating from the core are hot gas (plasma) that has been ejected at high speeds from the binary pair. The plasma emerges as a 400-billion kilometre-long geyser which is twisted by the force of the explosion and channelled upwards and outwards by strong magnetic fields. The flowing material appears to bend back on itself in a spiral pattern, perhaps due to obstruction in the path. (Courtesy of ESA, NASA and the Space Telescope Science Institute)

Borealis lies about one degree south of Epsilon Coronæ Borealis. This location is worth checking regularly by naked-eye observers. In a telescope T Coronæ Borealis usually appears as a tenth-magnitude red star, and it shows ellipsoidal variations with the orbital period of 228 days. These variations were first reported from an analysis of visual observations by members of the BAA.

CH CYGNI has been known since near the beginning of this century as a semi-regular variable, with a short cycle of 97 days superimposed on a 4,700-day change in the mean brightness. A series of outbursts began in 1966 which, together with spectroscopic changes, showed that it was a symbiotic star. The outbursts continued until 1984, when the star faded sharply. In the last few years the star has been fainter than ever, near ninth magnitude, and its semi-regular oscillations in a 97-day cycle have resumed. An explanation that has been put forward for the recent faintness of CH Cygni is that the hot companion that was the seat of the outbursts, together with its accretion disk, went into eclipse behind the red giant in 1984. The orbital period of CH Cygni is unknown, and it may be many years, but at the time of writing in early 1992 there was still no sign of any recovery, and a stellar eclipse lasting eight years or more would be unprecedented. If the hot star does emerge from eclipse, however, CH Cygni may rise quickly to near naked-eye brightness. Alternatively, the eruptive activity may simply have died down, but there is a good chance that it will start again. CH Cygni is a favourite star among binocular observers. Figure 2 is an identification chart for this star.

RS OPHIUCHI is another recurrent nova, with recorded outbursts in 1898, 1933, 1958, 1967 and 1985; others may have been missed. It was discovered in 1901 by Wilhemina Fleming at Harvard College Observatory, from examination of spectral plates taken in 1898. At minimum it shows erratic variations, usually between magnitudes 11 and 12, though subsidiary rises to magnitude 10 have been reported. The search for a period in the light variations, and for an orbital period in radial velocity measurements, has so far been inconclusive. Outbursts of RS Ophiuchi have been separated by as little as nine years, so the next outburst could take place around 1994. When it does come it will probably be found by an amateur astronomer. Observers can watch for the star about 4°N of Nu Ophiuchi.

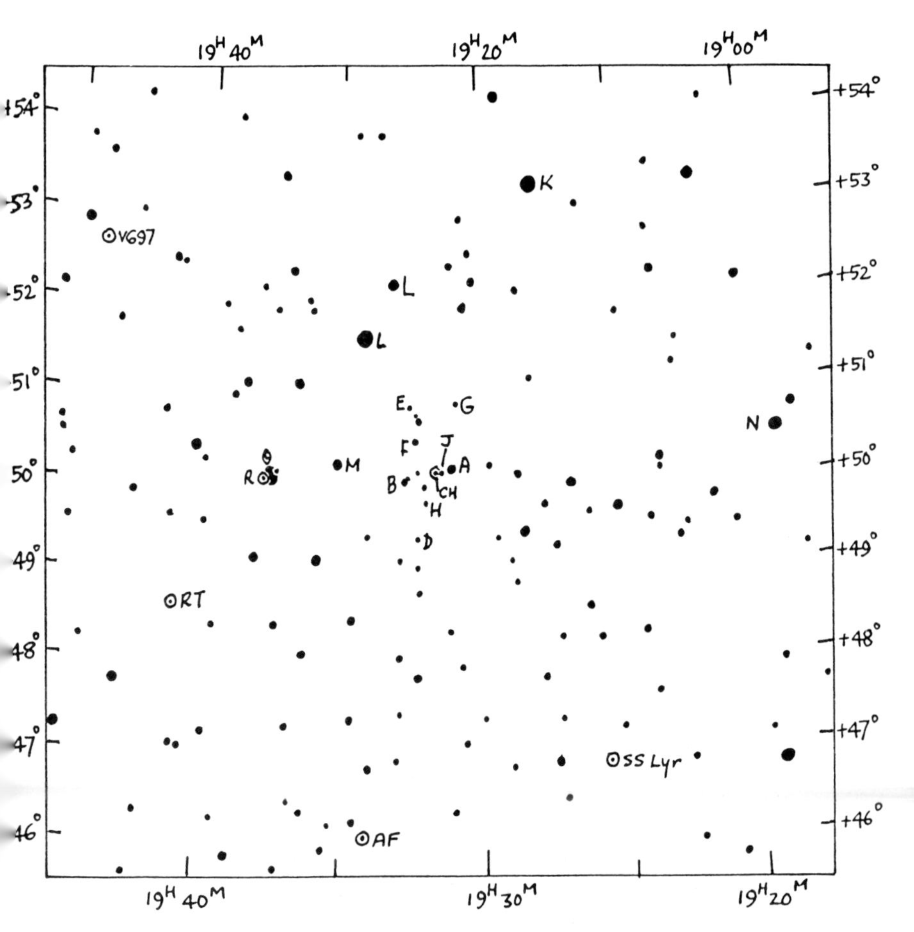

Figure 2. A chart for the symbiotic star CH Cygni. Comparison stars, with their visual magnitudes, are: N, 5.38; M, 5.53; L, 5.75; K, 5.91; A, 6.5; B, 7.4; D, 8.05; E, 8.1; F, 8.5; G, 8.5; H, 9.2; J, 9.4. The chart is adapted from one issued by the British Astronomical Association.

AG PEGASI is classified as a very slow nova, whose outburst began in the 1850s. Including the very slow decline, it lasted about 100 years. Unnoticed at the time, the outburst's course can be reconstructed from photographs and old star catalogues. AG Pegasi is still visible today as a star of about magnitude 8.5, 3.5° NE of Epsilon Pegasi. Its magnitude oscillates detectably as the system rotates in the orbital period of (most probably) 827 days. This seems to be because the side of the red giant that is irradiated by the hot star is brighter than the side that is in shadow. There is a nebula surrounding the system, detectable at radio wavelengths, comprising material that has been shed by the red giant and that is ionized by the hot star.

Observing symbiotic stars

As in many other branches of astronomy, amateurs play an important role in the study of symbiotic stars. Light curves showing the long-term behaviour of the brighter symbiotic stars are provided regularly to professional astronomers from the collected results of organizations such as the American Association of Variable Star Observers (AAVSO) and the Variable Star Sections of the British Astronomical Association (BAA) and the Royal Astronomical Society of New Zealand (RASNZ). Amateurs are also usually the first to report these objects' unusual behaviour, such as nova-like outbursts.

The visual magnitude of a symbiotic star can be determined in the same way as for any other kind of variable star: by comparison with one or more standard stars that do not vary. For readers who would like to identify a symbiotic star and record its magnitude (see Figure 2) which shows the comparison stars used by the BAA. CH Cygni is always above the horizon at the latitude of the British Isles, though it is low in the north on spring evenings. It can be located using the guide stars Iota and Kappa in Cygnus' wingtip, on the side nearer to Vega. To help you find these, you can refer to the Northern Star Charts (11L bottom) at the beginning of this volume, where Iota and Kappa are plotted but unlabelled.

While you may be lucky enough to find CH Cygni brighter than magnitude 6, bear in mind that it could be as faint as magnitude 9, which may be below the limit of your binoculars, especially if the field is low or if your skies are light-polluted. It is impossible to predict the star's behaviour, but in a good sky it is normally visible with 10 × 50 binoculars. When you have found CH Cygni, select a comparison star that appears equal in brightness to the variable, or interpolate between two of the comparisons to estimate the magnitude of CH Cygni to the nearest tenth. A graph of your magnitude estimates, made about once a week, may reveal its semi-regular pulsations with a period of about three months. Alternatively, perhaps you will detect the star in outburst.

Michael Victor Penston, 1943–1990

A remembrance

J. V. WALL

Michael Penston and I were brought together by quasars. How could the most distant objects in the Universe do this? When I started my professional life as an engineer in Canada and Michael his as a mathematician in Cambridge? When I went from Australia to the UK in 1974 and he went from the UK to Australia in the same year? We laughed about the long and tortuous connection more than once: can anyone doubt that quasars are the most powerful objects in the Universe?

We laughed a lot together; but to say that Michael and I disagreed on many things would be an understatement. For instance, we attacked our science quite differently. How do you expect to understand the Universe by looking at samples? he'd enquire. What are you doing binning everything up and looking at the statistics of apples and oranges? And what is the use of statistics anyway? Why don't you engineers/radio astronomers/wire-welders (whichever term he felt might provoke me more at the moment) learn how something works before trying to understand group behaviour? I barely ever dented his self-assurance with my retorts. How are things going in the pathology lab today, Michael? NGC 4151 behaving strangely still? What are you telling me about the Universe from your detailed study of a couple of its most paranoid members, Michael? How's the weather on NGC 1068? He was always ready and waiting for it, eyes glinting.

His eyes did not glint, however, when teaching or supervising was involved. At this point Michael became very serious. He regarded supervision of his students as of supreme importance; and one of his many legacies is the number of distinguished astronomers in the community fortunate enough to have been adopted as students by Michael and to have received their inspiration from him. His lectures were immaculate in preparation, delivered with customary style and wit; it all looked so easy. He recognized how important it was for astronomy to be taught and taught well as early as possible, and as a member of the Education Committee of the RAS he made

Michael V. Penston

Figure 1 (opposite). A montage of material left by Michael V. Penston on his office door.
'No Admittance' (but the Spanish have a more polite way of putting it), from his days as Observatory Controller, IUE tracking station at Vilspa, Madrid.
'MVP – Rookie of the Year' from his time in the USA (MVP = Most Valuable Player, as well as Michael's initials).
A cartoon about the missing socks; Michael's mismatched socks of psychaedelic colours were internationally famous. ('They aren't mismatched: they are H-alpha and [OIII].')
A further cartoon from a desk calendar, with a small adaptation of the caption to make it appropriate for the door of Michael's office.
Finally, a witticism of the type which so appealed to Michael.

Suddenly, through forces not yet fully understood, Dr Penston's office became the center of a new black hole.

RESERVADO DEHECHO ADMISION

I am neither FOR nor AGAINST apathy

a very substantial contribution to this. He was chairman of the Committee at the time of his death, a role which his wife Margaret has now taken up, and to which she is bringing her great experience and abundant energies.

Michael was always stimulating. Even his office door was stimulating. Figure 1 is a montage of some of the material as he left it there.

In addition to immense intellectual abilities (at physics, mathematics, astronomy, bridge, chess, whatever) Michael was all about putting things together in a simple way. (Perhaps it was all one and the same thing.) He explained them in simple ways with even simpler images. Remember Michael plus conker on a string on Patrick Moore's *Sky At Night* programme, demonstrating how the black hole in the nucleus of NGC 4151 had been weighed? A couple of examples of Michael's lucid cartoons to explain complicated matters are shown in Figures 2 and 3. The ideas seemed so simple; why hadn't we thought of them before; so simple that didn't we already know these things? No, is the answer. Michael had taken us there.

Part of Michael's stimulating nature made him take enormous delight in discomforting the Establishment or anything he regarded as the Establishment. Two little stories from the weighing of the Black Hole in NGC 4151 illustrate this.

1. Michael (and his colleagues) got the wrong answer, a simple arithmetical error which put the Black Hole too light by a factor of 10 (Monthly Notices of the Royal Astronomical Society, vol. 206, page 221, 1984). Michael took the (wrong) answer to Cambridge where he discussed it with the theorists. But we've always known that was the answer, said the theorists. The weight of the central Black Hole has always been 107^{8} solar masses. (In parentheses: why did you bother to invent this lovely technique and make all these observations?) When the right answer turned out to be an order of magnitude greater, his glee was enormous, barely tempered by concern at making a mistake (which was in fact very uncharacteristic).

2. The other source of glee was how the error was discovered. Was it picked up by his colleagues checking the manuscript before submission to the journal? No. By his department, in careful review? No (Gulp!). By the referees of the manuscript?

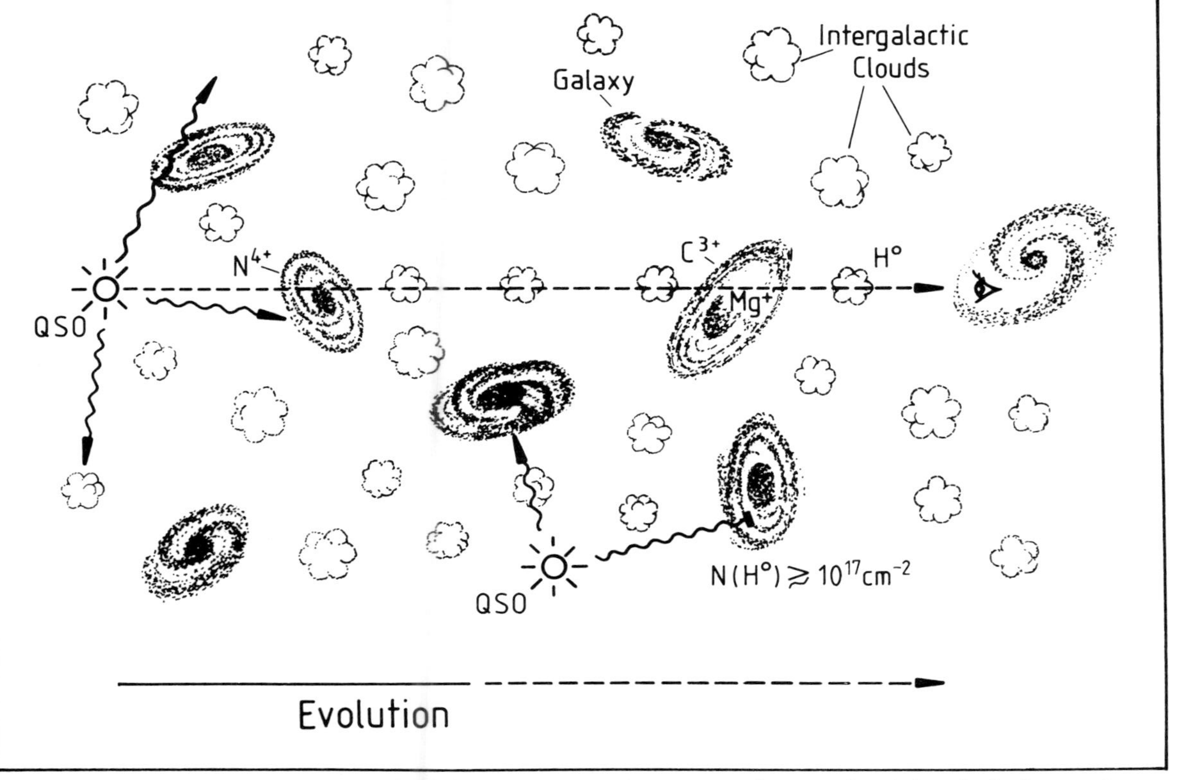

Figure 2. The spectra of quasars in the optical and ultraviolet are cut up by numerous absorption lines believed to be due to various objects between us and the quasar. Here is Michael's cartoon of the various causes of the absorption; the light from the quasar traversing space between us (Penston's eye in the Milky Way spiral galaxy) through other galaxies and intergalactic clouds along the line of sight.

No. By the editors of the journal? No. Then surely by the community of professional astronomers who carefully read, examine and dissect each other's learned publications? No. It was in fact found by an amateur astronomer in Switzerland, who wrote to ask why he couldn't get his arithmetic to come out right. Michael's joy was unbounded. It is the sort of little story which Michael honed by repetition to friends, colleagues, anyone in the circle who listened (and you couldn't not listen to Michael); and his infectious laugh chased up and down the corridors again and again.

Then there were his collaborations, constructed with characteristic direct simplicity and held together not so much by plans, dreams or visions but by example, dynamism and achievement. The last one

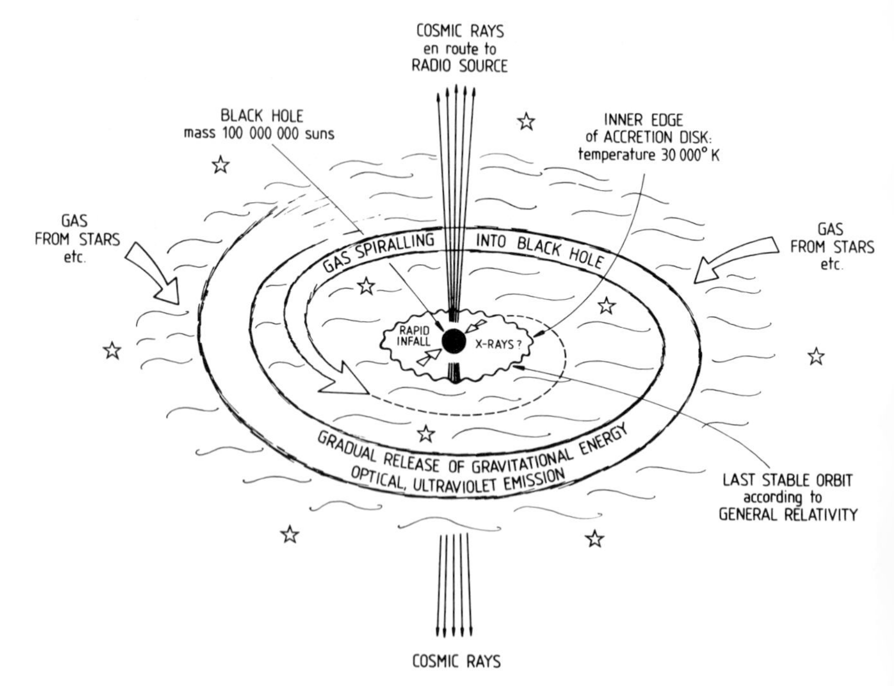

Figure 3. How does a quasar work? Penston's drawing of the innermost regions demonstrates the conversion of stellar fuel by gravity into radio, optical, and ultraviolet radiation.

was the biggest: over fifty astronomers joining in a consortium to win time on big telescopes in order to understand active galaxies. How could telescope-time assignment committees (the 'Establishment'?) resist. LAG it was called, the Lovers of Active Galaxies. Only Michael could have got away with such a name. LAG? Lag, yes, of course, I'm an old lag he would say.

We could have collaborated; we did indeed have our names together on a couple of papers. But this was more or less accidental, and we never got to write the big papers. In 1987 I went off to look after the Isaac Newton Group of Telescopes on La Palma. That's it then, he said. Six months. I give you six months of scientific life. Then you'll be out of it. You'll know only how to solve the 15-minute problems – fix it in 15 minutes or it can't be fixed. The deep thought about science will leave you, the concentration. . . . He was perhaps more than half right; I am fighting to prove him wrong. When I returned, he was ill, and he died in December of 1990 after a tremendous struggle against intestinal cancer. He was but forty-seven, still with his boyish enthusiasms, never to grow up. It seems hardly possible that he is not with us. A year later we still catch ourselves saying I dunno but I'll ask Michael – just a step down the corridor. We catch ourselves thinking of what Michael's opinion would be of this or that. And here is yet another lasting legacy: the ethos and the tradition of high principles, intellectual honesty, and scholarship which all of us who knew him must try to uphold.

But his style, his warmth, his stimulation, his enthusiasm and ebullience . . . his sheer exultation in knowing and understanding something about his Universe; where do we find this? A year on, and the loss feels greater than ever.

The 1990 Great White Spot of Saturn

MARK KIDGER

Introduction

The last twenty-five years have been sensational for planetary astronomers. Just over twenty-five years before these words are being written the first photographs had been taken from the surface of the Moon by the Soviet probe Luna 9. In the two and a half decades since then the Apollo programme has taken the study of the surface of the Moon from being one of mainly remote sensing to one of on-site study. Space probes have flown by and taken high resolution images of all the planets apart from Pluto (and even Pluto may be the target of a probe in the comparatively near future). Other probes have flown by the comets Giacobini–Zinner and Halley and, before this article is published, Grigg–Skjellerup will also be the target for a fly-by mission. There has even been a fly-by mission to an asteroid. An orbiter is on the way to Jupiter, albeit by the 'scenic route', taking in Venus, the Earth–Moon system and probably two asteroids en route. In the last twenty-five years we have found out more about our Solar System than in the previous three and a half centuries of telescopic observation for the Earth.

I said that they have been great years for planetary astronomers. Sadly though, for the amateur, they have been disheartening years indeed. Who in their right mind is going to spend cold hours in a dome in their garden trying to draw the fuzzy, shifting disk of Mars when the Viking probes have sent pictures from the surface? How many people want to freeze while scanning the disk of the Moon looking for TLPs when the seismometers left on the surface of the Moon by the Apollo astronauts have let us study the distribution of seismic activity on the lunar surface? Who wants to draw Jupiter when the Voyager probes have sent back pictures showing incredible detail in the clouds and revealing structures as small as a few kilometres in diameter on the surface of the five largest satellites of the planet?

It seems as if we know it all and, as a result, the number of active visual observers of the planets has fallen considerably since the mid-sixties. Over the last few years there has been something of a recovery, but many amateurs ask 'Why bother with the Moon and planets?' In this article I hope that I will show why there is still an ever-smaller, but still dedicated band of amateur planetary astronomers who still observe with their own telescopes or any instrument that they can get loaned to them. At a talk in Basingstoke, England in late 1991 I stated that there has never been a better time to start observing the planets, and I persist in this view.

During the last few years the visual observer of the planets has come back into his own and Earth-bound observation in general has picked up, both amateur and professional. A good barometer of the state of things is provided by the IAU Circulars; these record interesting and important observations and discoveries made by both amateurs and professionals in all fields of astronomy. In some ways, a scrutiny of past IAU circulars is one of the best ways of finding out which have been the most interesting and active fields of observational research over the years. To be published in an IAU circular the observations must be of considerable interest and justify rapid publication. In 1990 and 1991 no less than twenty-three IAU circulars carried details of planetary observations, ten of them giving details of amateur visual observations; the planets covered were Mars, Jupiter, and Saturn, whilst professional observations also covered Venus and, more surprisingly, Neptune. In the last few years the amateur visual observer has witnessed one intriguing phenomenon after another. The most spectacular though was the Great White Spot which appeared on Saturn in 1990. Although long-expected, it was amateur visual observers who discovered it and made the initial observations and amateur astronomers who persistently told the professional astronomers where to look. If proof were needed, when the journal *Nature* published an authoritative study of the 1991 Great White Spot of Saturn by the Spanish astronomer Dr Augustín Sánchez-Lavega, a renowned planetary expert, in the acknowledgements at the end of the article he gave due credit to the amateur visual observers who had supplied him with observations.

Amateur astronomers in Tenerife were in the thick of the observations of the Great White Spot, having made what are believed to be the first observations of it from outside the United States. At the time of writing (January 1992) the disk of Saturn is still active, at least it was when the last pre-conjunction observations were made in late November 1991, and the long-lasting effects of the great White Spot have still not faded away. Despite this, perhaps now is as good a time as any to give an overview of this spectacular event mainly from the point of view of the amateur visual observer who has, once again, demonstrated his value, even though he cannot compete with some of the instruments (such as the Hubble Space Telescope) which were rapidly directed at this remarkable phenomenon.

History

A Great White Spot is one of the most spectacular and also one of the rarest of large scale phenomena in the Solar System. The atmosphere of Saturn is much less active than that of Jupiter and, to an observer with a small telescope, usually presents an almost featureless disk, usually crossed by only a single, moderately prominent, but structureless, belt (the North or South Equatorial Belt dependent on which hemisphere of Saturn is presented towards the Earth). Occasionally though, for reasons that are not at all understood, a great eruption occurs at a deep level in the atmosphere of Saturn and, without warning, a giant and brilliant white spot bursts through the cloud cover. This phenomenon is unique to Saturn. Occasionally, white spots have appeared on Jupiter, one appeared at high latitude on Jupiter in 1990. These spots though tend to be very small and short lived; they cannot be compared to a Great White Spot on Saturn. Although I had seen drawings of Great White Spots on Saturn, nothing had prepared me for my first sight of the 1990 Great White Spot. There is nothing similar in the Solar System and the Great Red Spot of Jupiter pales by comparison.

We know of four Great White Spots in the northern hemisphere of Saturn, prior to that of 1990. Spots have also been observed in the southern hemisphere (such as the one observed in 1969), but they are generally less prominent than those of the northern hemisphere. Of large scale phenomena on the planets, this is probably the least frequent, apart from the occasional disappearances of the SEB of Jupiter. When they have appeared, the brief duration of the spots has made it very difficult to study them in detail.

Despite the fact that over the centuries many great observers such as Cassini, Huygens, William Herschel, Schröter and Lassell have observed Saturn and made significant observational discoveries, the first Great White Spot (GWS) to be observed was that of 1876. This spot appeared in the EZ (Equatorial Zone), slightly north of the equator. The discovery was made by Asaph Hall, on December 7, 1876, shortly before discovering the satellites of Mars.

This was the least observed of the five GWSs which have been seen; the last of the sixteen known observations was made thirty days later, on January 6, 1877. One wonders though, was the lack of observations of previous GWSs simply due to the fact that comparatively few visual observations were made of Saturn and that the spots were simply missed? The 1990 GWS could be and was seen, even by inexperienced observers, through a 60-mm refractor and was easy in any aperture of 80 mm or greater. It is hard to believe that there were no GWSs prior to that of 1876 and almost as hard to believe that such a spectacular phenomenon could go undiscovered for two centuries of detailed telescopic observation of the planet.

The following GWS to that of 1876 also had a famous discoverer. On June 15, 1903, Edward Emerson Barnard discovered it in the NTZ (North Temperate Zone). Amongst other observers of this spot the name of José Comas Solá stands out, probably the greatest Spanish observational astronomer of all time, precedent to the large contribution that Spanish amateurs made to the 1990 observing campaign. The Barnard GWS split in two shortly after appearing and, five and a half weeks after the original outburst, a third spot appeared. The total duration of the three spots was 150 days.

Without doubt though, the most famous GWS on Saturn was that of Will Hay in 1933. Hay was not an illustrious professional astronomer like Hall and Barnard, in fact, he was a comic actor, well known in the 1930s and '40s for his performances in films, some of which are still aired occasionally on television. Apart from being a talented actor, he was an enthusiastic observer, who found the GWS on August 3, 1933, whilst observing with his own 15-cm telescope. Curiously, he saw no great merit in his own discovery and even wrote to the *Journal of the British Astronomical Association* apologizing for having made the discovery himself instead of some other, more worthy observer. Once again, the spot appeared in the E2, slightly north of the equator. As in 1903, several spots appeared in all following the original one; in fact, a total of four GWSs were observed in total, the last appearing on September 27, 1933. In all, the spot lasted a total of nearly fifty days. During the first three weeks of observation, it became more extended, growing from 25,000 km in length on August 3, to 69,000 km on August 22.

Least prominent was the 1960 GWS, the last one to be observed before that of 1990. The 1960 GWS was discovered by Botham (not, I am given to understand, the same Botham as the one who plays cricket for England!; in fact, I have found no biographical information at all about him) on March 31, on the border of the NTZ (North Temperate Zone) with the NPR (North Polar Region). The first spot was only seen for forty days, although three others appeared over a total of six months.

Precedents to the 1990 GWS

Looking at the years 1876, 1903, 1933, and 1960, we see a separation of 27, 30, and 27 years between appearances. The mean interval is very close to Saturn's orbital period (29.5 years). For this reason, various astrophysicists such as Ignacio Ferrín and Augustín Sánchez-Lavega predicted that a similar spot should appear approximately between 1988 and 1990. During the last two years a certain nervousness has appeared in the astronomical community to see if Saturn would comply with the predictions. The 1990 opposition of the planet occurred on July 14, without the slightest indication of abnormal activity (as we will see, this may well not have been true). Some people even speculated that the spot might have appeared during the two 'blind' months around conjunction.

However, even in the observations taken in 1989, there are indications of activity starting to develop in the disk of the planet. Figure 1 shows a graphical representation of intensity estimates of the EZ made by members of the Agrupación Astronómica de Tenerife between May 20 and September 17, 1989. Despite the dispersion in the data (logical given that this was the group's first serious campaign of planetary observation), a significant increase in the brightness of the EZ was registered. At the same time, a smaller increase was seen in the brightness of the NTrZ (North Tropical Zone), whilst the brightness of the NTZ remained constant.

Some time ago, I became aware of a peculiar symmetry in the observations of the known GWSs on Saturn which, as far as I can tell, I have been the first person to draw attention to. We see what is, at very least, a curious coincidence in the positions of and intervals between the five known GWSs. The intervals between apparitions are: 26.5, 30.1, 26.7, and 30.5 years, whilst the spots have appeared successively in the EZ, NTZ, EZ, NTZ (in fact, this GWS appeared in the NNTZ, which does not really affect my argument) and EZ. There seems to be a pattern in both sequences. There is a strong temptation to predict that the next GWS will appear in the NTZ in spring 2016 and will be considerably less spectacular than the astonishing GWSs seen in 1933 and 1990.

The mean interval between the apparition of the five spots has

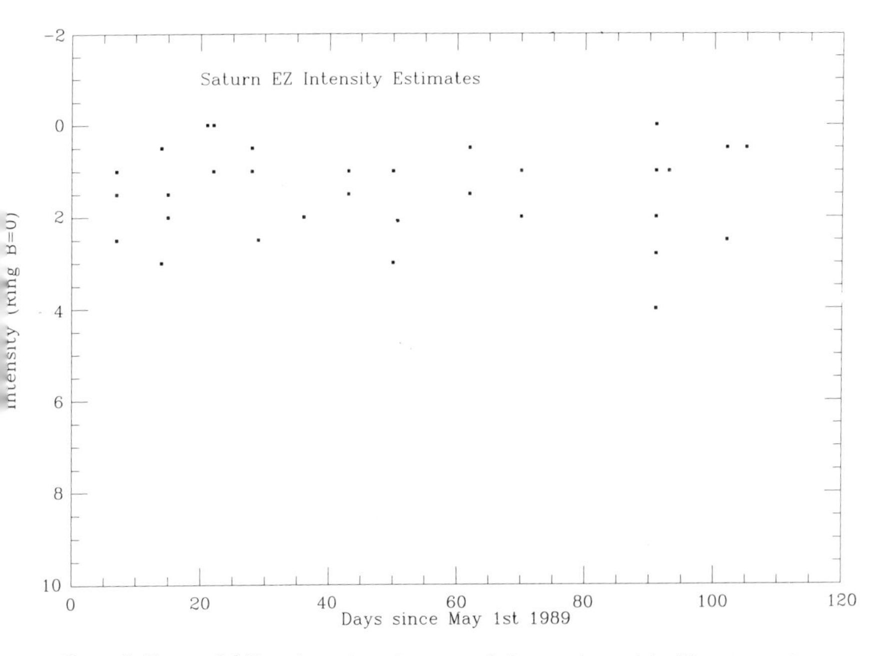

Figure 1. Equatorial Zone intensity estimates made by members of the Planetary and Lunar Section of the Agrupación Astronómica de Tenerife in 1989. A least squares fit shows a clear increase in the brightness over the summer. This is also reflected in the monthly averages.

been 28.5±1.9 years. Successive spots are almost a constant amount either ahead of, or delayed with respect to, the average interval between them. It is hard to imagine that this could be due to anything other than coincidence and, were it to be found to be other than coincidence, even harder to imagine what conceivable physical reason could exist to explain it, but this detail is yet another peculiar aspect of the history of these spots.

With five GWSs now known, the mean interval between their appearance is seen to be significantly less than the 29.5 year orbital period of Saturn, although the difference is considerably less than the dispersion in the intervals between appearances. Augustín Sánchez-Lavega, someone far more expert in these matters than I am, persists in his view that there is a strong tie-in with the 29.5 year orbital period – and yet I wonder.

The observation of the 1990 GWS

Unfortunately, we were not able to see the 1990 GWS in the most favourable circumstances, as Saturn was at opposition on July 14, more than two months before the appearance of the spot. Although this meant that Saturn was an evening object (more favourable for observers) whilst the spot was on the disk, its period of visibility was reduced night by night during the two months that it was observed; worse still, Saturn effectively disappeared from view at conjunction whilst there was still a considerable amount of activity on the disk and new white spots were still appearing.

The consequence of the appearance of the GWS occurring well after opposition was to impose an ever-narrower window during which observations could be made. Given that Saturn's rotation period is approximately ten and a quarter hours, the same longitude is on the meridian and thus exactly the same face of the disk is presented to us approximately fifteen minutes earlier every third night. At the start of October, for most observers, this meant that the GWS was on the meridian during the observing window on two nights in every three. By mid-November, this had reduced to once every three nights maximum, whilst, for many observers, no meridian crossing occurred during their very brief observational window before Saturn became too low in the sky to observe. Even for those observers able to see the GWS during November, the conditions were difficult as Saturn was at a very southerly declination and thus difficult to see from northern latitudes. As conjunction approached, observers had to contend with steadily worsening seeing due to the low elevation of the planet in the sky as well as an ever-briefer interval in which to make their observations.

Our knowledge of the 1990 GWS is thus incomplete in the sense that observations could be made for just two months and then, for another two months, Saturn was effectively hidden behind the Sun and unobservable. We know though that, in the case of the 1960 GWS, new nuclei of activity appeared for six months after discovery, whilst the 1903 GWS was also active for some six months. It is as if we have a book with the middle chapters removed and the preceding ones defaced: we can read the first two chapters complete, the third and fourth are only partially legible and then there is a gap to the last chapter, by which time most of the action has finished and only the loose ends remain to be tied up.

Comparison of the available observations of the current GWS against the evolution of previous spots shows that the 1990 spot has been exceptional. Both in its rapidity of evolution and in the number of individual eruptions, the Wilber GWS appears to have been more active than any other. Once again, referring back to my suggestion that there is a symmetry in the intervals between appearance and location on the disk of the different GWSs which have been observed, it seems that the three spots which have appeared in the Equatorial Zone have been the shortest lived and most rapidly evolving. The more we compare the observations, the more remarkable is the split of the observed spots into two groups, an 'equatorial group' and a 'temperate group', with great similarities between the spots within each of the two groups, but very different characteristics between the two. It is a pity that, statistically speaking, our sample is so small, although it will be of exceptional interest to observe the next GWS and see if it really is more similar to the longer enduring, but less active and less spectacular polar spots of 1906 and 1960.

The rapid evolution and considerable activity of the 1990 Wilber spot, as well as its limited visibility, has caused considerable difficulties when analysing the observations. The original spot only remained visible as a spot for a few days, evolving at an ever-increasing pace. With the successive eruptions, it was the observers who often had difficulties: when timing a meridian pass they were often not sure exactly which of the different nuclei they were observing and at least one possible eruption is unconfirmed because of confusion between the different individual nuclei. If this were not sufficient, in much of Europe, the weather was very poor during the weeks following the appearance of the GWS. Especially seriously affected were the different groups in Catalonia (Spain), a region which has a long tradition of amateur astronomy and whose observers are normally in the vanguard of European planetary observations.

Overview of activity

The first observation of the 1990 GWS was made by Stuart Wilber, an American amateur, at approximately 05.00 UT on September 25; various astronomers from Las Cruces confirmed the observation and thus ended an anxious wait for the spot. As frequently happens in the case of such phenomena, many other observers picked up the spot independently before reading of the official discovery. As is usually the case, there is also at least one observer who makes the discovery completely independently at the same time as, or ahead of, the 'official' discovery but either does not report his observation, or reports it very late. On this occasion it was a Californian amateur called Alberto Montalvo, who picked up the GWS on the 25th, but only reported it three days later. At the next favourable meridian crossing, three days later, various American visual observers confirmed it. The first known European observation was made by myself at 20.00 UT on September 29. Due to the observing geometry referred to previously, assuming that the eruption actually occurred within one day of Stuart Wilber's first observations, the GWS could have been picked up from Japan on September 24 or 27, from Europe on September 26 or 29 or from the USA on September 25 or 28. Had it been cloudy on the west coast of the USA on the night of discovery, it could easily have been picked up by the highly active Japanese astronomical community on the following night; in the end, the American observers beat everyone to the draw, but luck, as always, played a significant part in the discovery.

My own first European observation was an example of how much chance influences in such matters. The Agrupación Astronómica de Tenerife had been observing Saturn for two years in the hope that a Great White Spot might appear. During the summer of 1990, due to my other commitments, I had not gone up to the telescope with them for some six weeks and had specifically requested that they might pencil me in for the night of September 29 as I wanted to get an observation. At exactly 9 p.m. (20.00 UT) Saturn entered the field of view of the 10-cm finderscope of the 51-cm reflector. I looked at the disk as I was centring the telescope and thought that I could see a bright white spot to one side of the meridian. Without saying anything to the four other observers, I quickly centred the planet in the main instrument and focused up, thinking that my imagination (or wishful thinking) was playing tricks on me. When the same white spot was visible through the ×270 eyepiece of the telescope my next thought was 'for heaven's sake make sure that someone else looks quickly and confirms that I am not dreaming'. I think that it was Eva Redondo-Terrón who, not quite understanding my frantic excitement, took over at the eyepiece and confirmed that it really was there. For twenty-four hours I had the considerable illusion of thinking that I might have actually made the discovery; Alan Heath, British Astronomical Association Saturn Section Director tells me that, for the same period of time, he was even calling it 'the Kidger Spot': the disappointment on reading IAU Circular #5105 was considerable.

In various previous GWSs activity has been seen from one or several nuclei which have appeared after the first. The time scale has been anything from a few weeks to several months. In the case of the 1990 Wilber GWS, no less than four (and probably five) additional nuclei have appeared after the original eruption. On October 10 (15.6 days after the Wilber discovery), three new nuclei of activity were observed from Tenerife superimposed on the original spot, two of them extremely prominent. At the time of the original observation by Wilber, the GWS was at longitude (System I) 311°. The new nuclei (A, B, and C) were found at longitudes of A=314°, B=337°, and C=349° respectively. The coincidence in positions between nucleus A and the original eruption does not appear to be a coincidence, rather it may indicate the presence of an active zone in the deep atmosphere of Saturn, which has given rise to two separate eruptions. The fourth of the subsequent nuclei of activity (here denominated 'the Torrell Spot'), was officially discovered on November 2 by Sebastià Torrell. Some time after its announcement an observation reached me from Daniel Verde observing from Gran Canaria who had seen a bright spot at longitude approximately 140–145°. The longitude of the spot discovered by Torrell was 148°, and its drift in longitude established by later observations was small and positive, thus it seems certain that the two structures were the same, although it seems odd that nobody reported seeing it between October 20 and November 2. When asked why he had not reported his observation earlier Daniel said that he thought that he had been observing one of the previous nuclei, thus he missed the opportunity to claim the discovery for himself. The longitude of this spot was almost exactly 180° ahead of the initial GWS. Incredibly, every single one of the observed spots was either discovered, or independently discovered, by members of the Planetary and Lunar Section of the Agrupación Astronómica de Tenerife.

Some astronomers have suggested that these were not genuine spots but rather simple condensations in the original Wilber GWS. I beg to differ although it is obvious that the Wilber GWS has been different to any other GWS which has been observed previously. Two of the three nuclei which appeared on October 10 were incredibly intense and very much brighter than the Wilber spot on which they were superimposed and also very small and condensed:

it is very hard to think of them as condensations as their aspect was certainly not that of a simple condensation.

It is difficult to speak of the duration of activity of the different nuclei, partly for reasons explained below. However, in no case, apart from the Torrell spot, has the spot remained visible much more than two weeks. Although it was still possible to see the effects of the Wilber GWS after October 10 it now occupied nearly half the circumference of the EZ and could no longer be regarded as a spot, rather it had converted itself into a major disturbance of the Equatorial Zone. Different estimates state that it had spread around the entire equator of the planet between about October 22 and 24 although there do not appear to be observations which show the exact point when this occurred and can thus decide between the conflicting estimates. The Hay GWS though does not seem to have shown continued growth after about two weeks from discovery and the Hall GWS of 1876, although followed for forty days, was not noted to have grown at all. What started then as a simple spot seems to have evolved into one of the largest storms ever seen in the Solar System. Observations show that this disturbance lasted well into 1991; however, nuclei A, B, and C seemed to have lasted less than two weeks, whilst the Torrell spot was observed for at least five weeks, hence it was, by far, the longest lived of all the spots which appeared during this outburst of activity.

Size, expansion velocity and rotation period

The most important aspect of the observation of GWSs in Saturn's atmosphere is the enormous amount of information which it can give about the atmospheric circulation. Despite generations of effort by visual observers our information about the atmospheric rotation of Saturn is pathetically limited. The disk is divided into System I (the equator) with a rotation period of $10^h14^m08^s$, although different sources give slightly different values and some authorities round this value to exactly 10^h14^m, and System III (high latitudes) with a 10^h40^m rotation. Unfortunately, unlike Jupiter, the number of atmospheric details which can be followed over periods of time sufficiently long to get a reliable rotation period is very small; at high latitudes the situation gets even worse. No structure at latitude greater than 58° has ever had its rotation period calculated. In fact, according to Sánchez-Lavega, only twenty-three structures observed between 1793 and 1983 have had reliable rotation periods calculated for them; of these, four were GWSs. A Great White Spot provides a good, prominent structure in the planet for which meridian crossing times can be easily and accurately calculated. The fact that they are so prominent permits observations to be made with comparatively small apertures. A reasonable base of meridian crossing times permits a good rotation period to be calculated.

The factor which has most complicated the calculation of the rotation period of the Wilber GWS has been the rapidity of its lateral growth. We have seen that other GWSs (especially the Hay spot and probably the Barnard spot) grew over a period of several weeks. The Wilber GWS grew though, on average, approximately three times faster than that of Hay. Figure 2 shows the sizes of the Hay and Wilber spots against time. We can see that, initially, the two spots grew at the same rate. The Wilber spot though seems to have maintained its rate of growth longer than the Hay spot, hence the average rate of growth is faster for the Wilber GWS.

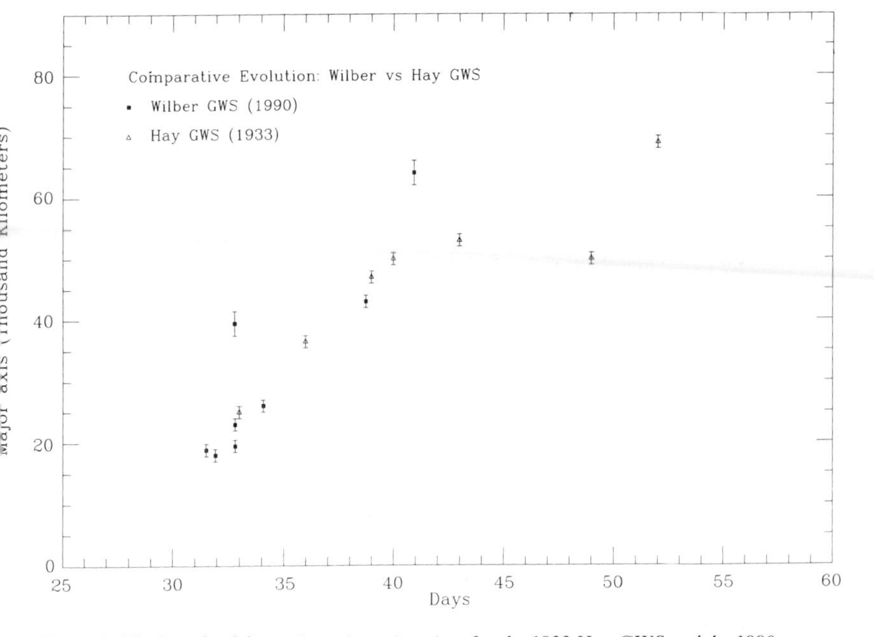

Figure 2. The length of the major axis against time for the 1933 Hay GWS and the 1990 Wilber GWS. The initial rates of growth are very similar, but the Wilber spot appears to maintain its rate of growth for a longer period of time.

The sizes of the GWS measured by meridian passes are presented in Table 1. The measurement technique consists of timing how long it takes for the spot to cross the meridian completely. Alternatively, one can use the meridian crossing of the centre and one edge (which, logically, gives the semi-major axis). Given the rotation period, which for System I converts to 844.3 degrees per day, the size of the spot can be calculated easily.

TABLE 1

The size of the Wilber GWS calculated from visual meridian crossing observations.

Date	*Size (km)*	*Observer(s)*
01-10-90	18,800	Miyazaki
01-10-90	18,000	Víctor González/Francisco Hernández
01-10-90	18,000	Mark Kidger/Jorge González
02-10-90	19,500	Víctor González
02-10-90	23,000	Francisco Hernández
02-10-90	39,000	Josep M Gómez
04-10-90	26,000	Steve O'Meara
08-10-90	43,000	Rob Moseley
10-10-90	64,000	Mark Kidger

The observations are seen to follow a well-defined progression of size with time from the eruption. One single point is seen to differ from the general trend; when this point is neglected the remaining values take the form of a logarithmic relation which is very accurately fitted by the formula:

$$1 = 10^{0.057d+4.16}$$

where:
1 = length in kilometres
d = date of the observation (days since October 0).

Effectively, the Wilber GWS grew by some 14 per cent per day. In other words, its length doubled every 5.3 days. This implies that, if this rate of growth were maintained, by October 24.8, the spot must have occupied the entire circumference of the planet. The estimates that the GWS actually completed its circumnavigation of the equator on October 22 implies that, if this were true, the rate of growth must actually have accelerated with time. Extrapolating the relation back in time, on September 25.2 the GWS could have been no larger than about 7700 km in length. It does not seem though that any estimates of the size exist in the early phase of the spot's development. The first known estimate of the size is one of 23,000 km taken from my own drawing on the 29th. This though is probably misleading as the spot was fairly close to the limb, the leading edge was very fuzzy and indistinct, with a well-defined edge marking the border of the inner core and then a far more nebulous extension to the west which later developed into the outflow of effluvium so spectacularly observed by many Spanish amateurs over the next week. Other observers who made rapid sketches on the same night (Eva Redondo-Terrón, David Macía and Nicolas González) tended to draw the spot rather smaller and in much better agreement with the expected size.

The average expansion velocity between discovery and extending round the entire disk, was 150 m/s; however, as we have seen, the rate of growth of the spot was exponential and the rate of expansion increased by some 14 per cent per day. Hence, in the initial phases, the expansion velocity of the GWS was very much lower, as is shown graphically in Figure 5. For example, in the twenty-four hours after discovery, the mean expansion velocity was just 12.2 m/s; this though had increased to 88 m/s by the time the wave of activity had started on October 10 and, in the last twenty-four hours before the two ends of the spot joined up, the expansion velocity had increased to some 550 m/s.

On the other hand, the average rate of growth of the Hay GWS was only 4 per cent per day; between a quarter and a third of the rate of increase of the Wilber GWS. The Wilber GWS has clearly been exceptional in its very rapid and sustained rate of growth. However, there are not a lot of details available as to this aspect of the evolution of previous GWSs, although in no case is there evidence of an exceptional rate of growth.

There is, though, the possibility of what has been termed jitter in the position of the GWS. On October 2 the observed meridian pass was delayed by some fifteen minutes compared to the predicted time, according to two independent meridian transit observations which were in perfect agreement, although the size appeared to continue its steady progression of increase. The one very discrepant estimate of the size of the GWS occurred on this date and appears to have been due to using an observed transit of the following edge of the spot along with the predicted time of the meridian pass of the centre of the spot. This delay translates into a very large temporal increase in the rotation period. This meridian pass is one of the best observed as there are no less than four observations of the transit of the following edge of the spot, which have a dispersion of just four minutes between them.

We can only speculate as to why the Wilber spot developed so rapidly. Though there are various clues which suggest that its nature was different to that of previous GWSs. When discovered, it was at longitude 315° (System I). Over the interval from September 28 to the last measured meridian pass on October 8, the longitude showed a steady drift with respect to System I. The rate of drift is shown in Figures 6 and 7 and amounted to an astonishing 3.1° per day. That's to say, the position of the spot lagged by some 1380 km per day or, alternatively, by some 58 km/h, or 16 m/s, compared to the planetary System I. The drift is very large compared to that measured for the two previous GWSs in the EZ. There is a large

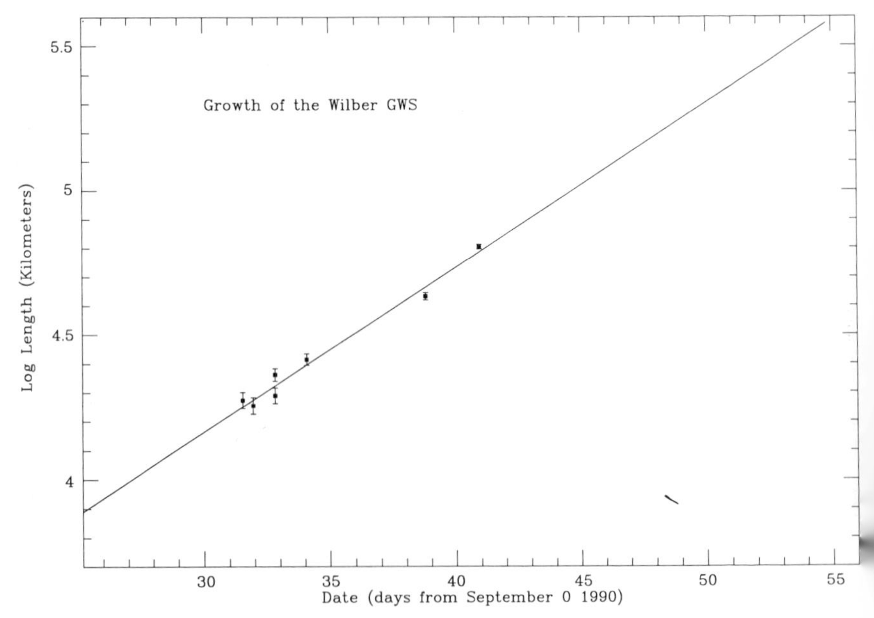

Figure 3. The growth of the 1990 Wilber GWS. The logarithm of the length of the major axis is plotted against time. The straight line shows that the percentage increase in size per day remains constant, thus the spot did not expand at a constant velocity.

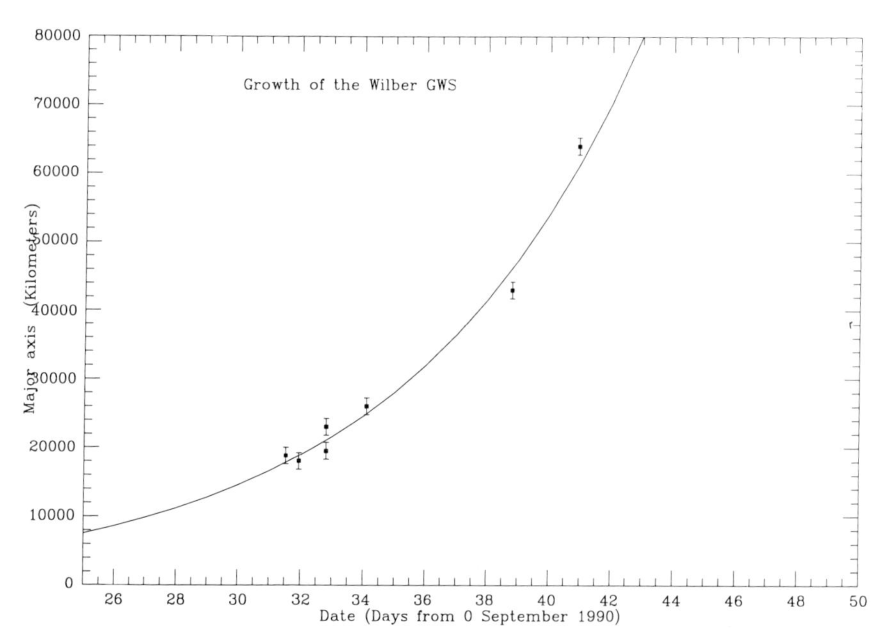

Figure 4. The growth of the Wilber Great White Spot expressed in linear units of kilometres against time. The exponential nature of the growth is very obvious.

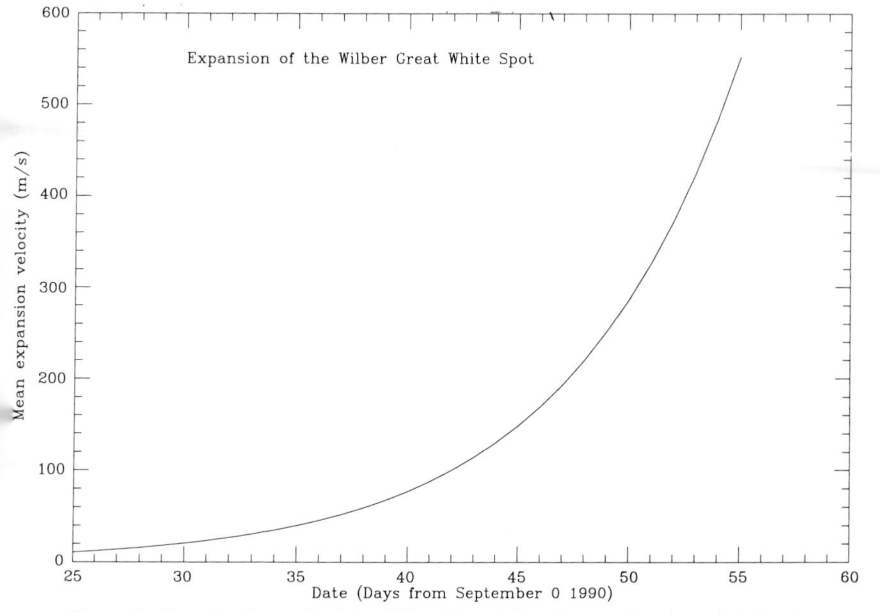

Figure 5. The rate of growth of the Wilber Great White Spot against time. In this plot the expansion velocity (in metres/second) is given against the date.

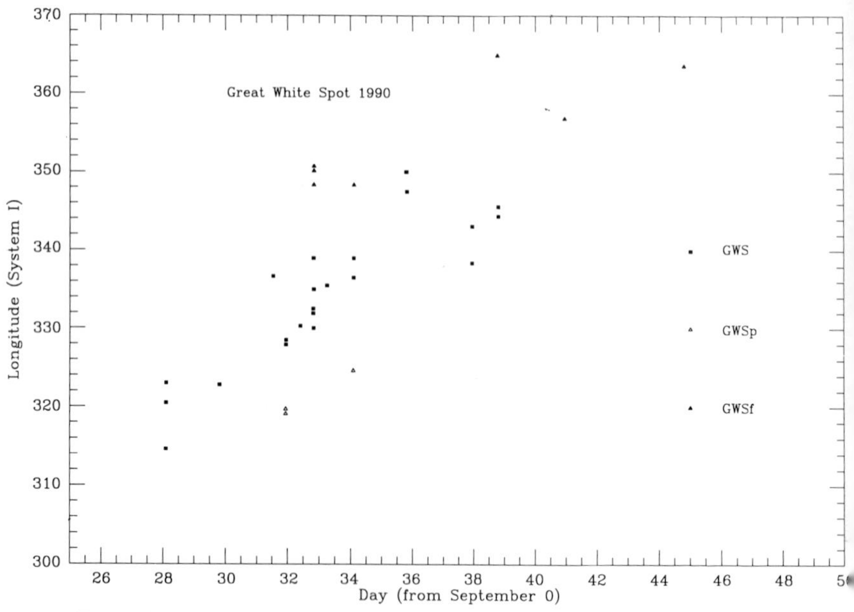

Figure 6. The position of the different nuclei of the Great White Spot in terms of longitude in System I. The large period change at around October 6.2 is manifested by a considerable change in the slope of the line representing the position of the centre and preceding edge of the spot.

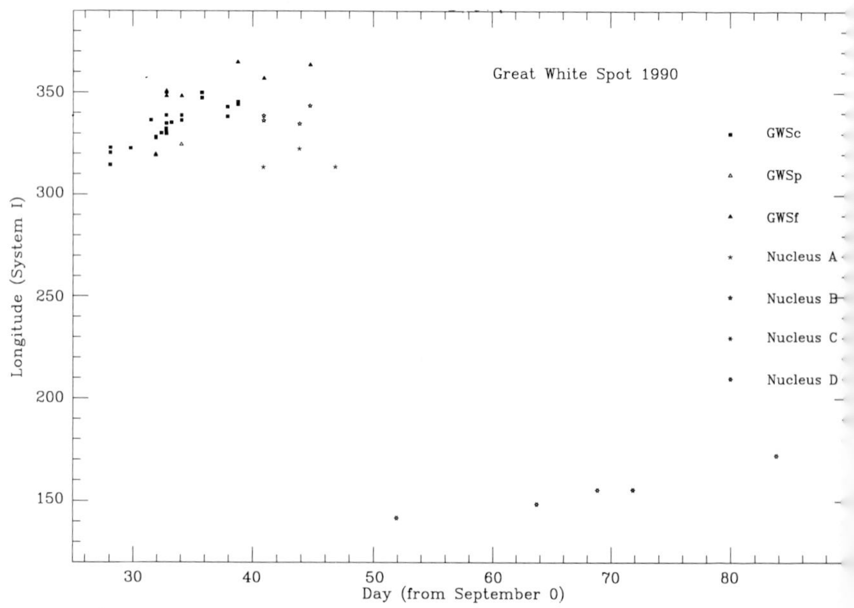

Figure 7. As Figure 6 for just observations of the Wilber Great White Spot.

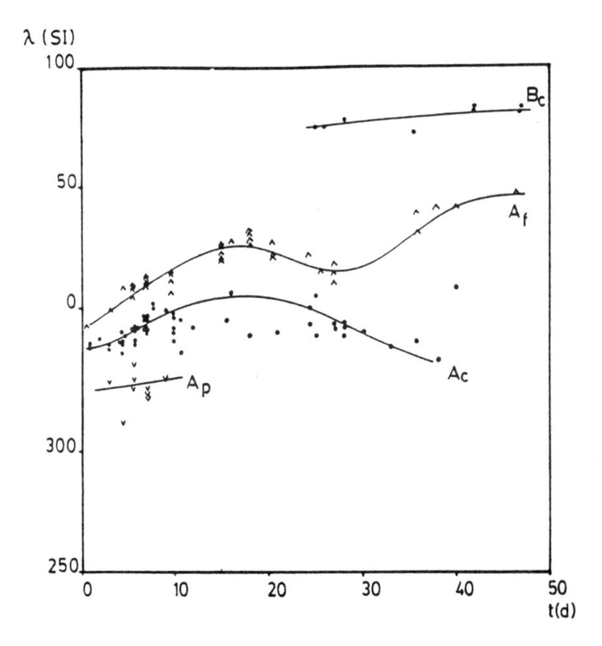

Figure 8. The position of the Hay GWS of 1933, the most similar in behaviour of previous Great White Spots to the Wilber Great White Spot of 1990. The drift in longitude of the centre of the spot is very small. This diagram is reproduced from the Doctoral Thesis of Augustín Sánchez-Lavega (Faculty of Sciences, University of the Basque Country, 1986).

archive of data of the Hay GWS which shows that it barely showed any drift at all in longitude. There is, though, sufficient dispersion in the data for the 1933 GWS that one can only say that, over forty days, it amounted to no more than 10° (see Fig. 8). The 1990 GWS seems, in addition, to have been the only one to have shown such sustained lateral growth; it also seems to have been the only GWS to have drifted so far and so fast in longitude.

One possible explanation is that the Wilber spot formed at a different level in the atmosphere to previous GWSs. Certainly there is some evidence that the Hay GWS was at a much higher level of the atmosphere than that of Wilber; this will be discussed later in this article. One possible explanation for both the drift and the very high rate of expansion is the presence of a Saturnian jet stream at this level, which, if it showed strong variations of velocity with height, would have the result of sheering the spot. Sánchez-Lavega finds that the cloud systems related to the disturbance in the Equatorial Zone gave zonal wind velocities of up to 450 m/s.

The rotation period of the different nuclei has been calculated from all available visual meridian transit observations made by the observers listed above. The largest contribution to this data base has been the observations made by the Agrupación Astronómica de Tenerife and, especially, the team of Víctor González and Francisco Hernández. However, the shortage of observations of some of the nuclei has caused large uncertainties in the calculations. There is also the possibility of temporary variations in the rotation period. Figure 9 shows the dates of observed meridian crossings against number of rotations of the planet, counting the first observed meridian pass as rotation number 1. The period is simply the slope of this graph. Note that the different lines represented by the meridian transits of the centre and two edges of the GWS are so close together on the graph that they cannot be separated.

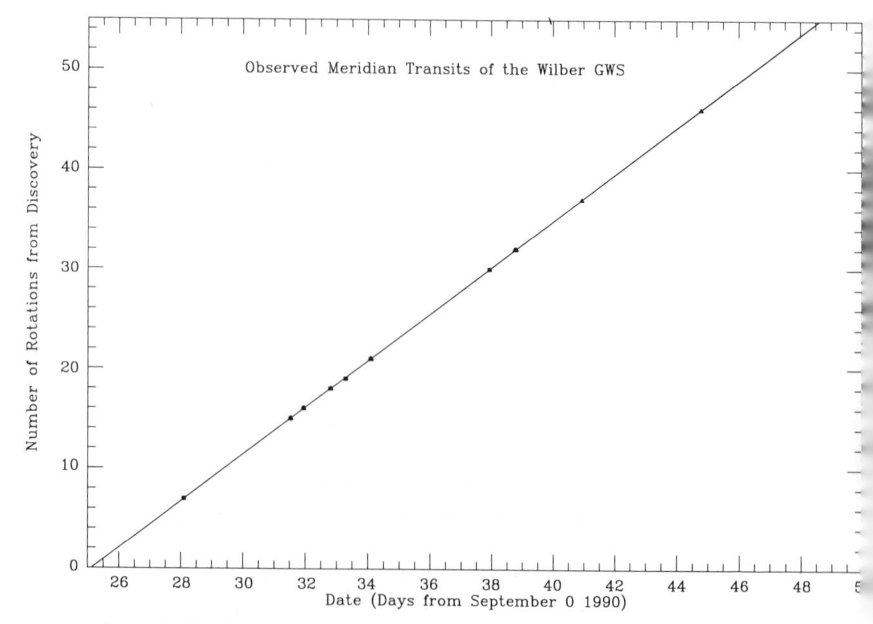

Figure 9. Meridian crossing times against number of rotations since discovery. The slope of this plot gives the rotation period for the GWS.

The calculated periods are synodic (meridian crossing to meridian crossing). For the Wilber spot the average values are:

GWSp	(leading edge)	—$10^h13^m56^s$
GWS	(centre)	—$10^h15^m35.9^s$
GWSf	(trailing edge)	—$10^h15^m14^s$

The intervals between the first and last measurement of each were 17, 25, and 39 rotations respectively. From these figures it seems that the mean period of the trailing edge of the GWS is less than that of the centre. However, as we will see, this is due to the fact that the data do not cover exactly the same interval of time.

The rotation period of System I is $10^h14^m08^s$ (0.426388 days), thus, as shown by the drift in longitude, the period of the GWS is larger than that of System I by 1^m28^s. However, the true situation is more complicated. Figure 6 presents the position of the spot in terms of longitude, referred to the Saturnian System I. We can see clearly how the longitude increased rapidly until reaching a maximum value in the first week of October. After this date, the longitude of the spot decreased rapidly. Separating the data into two independent linear regressions and solving for the date of intersection, we find that the longitude increased by 3.70° per day (i.e.: it was delayed compared to the EZ) until October 6.26, reaching a maximum of 347.6°. From this date, the longitude decreased by 2.0° per day. The drift of −3.7 degrees per day implies that the rotation period of the centre of the GWS was 0.4 per cent greater than the nominal System I rotation period of the planet.

These values imply a large change in the rotation period. Separating the observations into two groups once again, the periods before and after October 6.2 are found to be:

Before October 6.2	$10^h16^m42.1^s$
After October 6.2	$10^h12^m32.7^s$

That's to say, the period decreased by more than four minutes. The change in atmospheric circulation implied by this decrease is very large: a delay of 1650 km per day in the position of the spot suddenly changed to an advance of 890 km per day; a change from −19 m/s to +11 m/s against System I. These data again suggest that the GWS may have formed in a level of the atmosphere with a very large gradient of wind velocity.

Both the trailing and, presumably, the leading edge of the spot also showed significant changes in their period. Even though the mean period of the trailing edge is $10^h15^m14.0^s$, we find that its period decreased by several minutes at the same time as that of the centre of the spot, hence the result above that its mean period is less than that of the centre of the spot; this is simply due to the fact that there are meridian transits for the trailing edge which post-date those of the centre by nearly a week and thus heavily weight the mean period to the shorter value. This we see from the results for the epochs before and after the date of the sudden period change:

Before October 6.2, $10^h17^m11.4^s$
After October 6.2, $10^h14^m56.1^s$

In this case the change in the rotation period is not as large as for the centre of the spot, but it is still considerable.

The rotation periods of the four subsequent nuclei of activity appear to have been less than that of the original GWS, although the data are too few for this conclusion to be accepted as absolutely solid. There is only one meridian pass measured for nucleus C, thus we can say nothing about its rotation. Nucleus B does though appear to move with System I or, at least, have a very similar rotation period, although the rotation period of nucleus A is delayed by a similar amount with respect to the System I rotation period as the original period of the Wilber GWS. In fact, within the probable errors, the Wilber spot and nucleus A have identical rotation periods:

$$A = 10^h17^m06^s$$
$$B = 10^h14^m33^s$$

The fact that nucleus A had a rotation period almost identical to the original period of the GWS and formed at very nearly the same longitude (i.e. position in the atmosphere) appears to be significant.

The Torrell spot was still observable at conjunction although, by the time that Saturn re-emerged in early March, it had now disappeared. This spot also showed a pronounced drift in longitude with respect to System I, amounting to +0.95° per day. The provisional rotation period from October 20 to November 23 was:

$$\text{Torrell} = 10^h14^m56.8^s$$

This period of rotation, like that of the Wilber spot, is rather slower than System I, this time by 49^s. However, even this difference in rotation period is large compared to that measured for previous Great White Spots of Saturn. The time base for the observations of the Torrell spot is sufficiently long and consistent for this period to be accurate to within a few seconds. In contrast to the Wilber spot, there is no evidence at all of any period change in this object.

Table 2 shows the measurements of the size of the Torrell spot. The rate of growth was clearly much less than that of the Wilber spot, being just 6% per day from October 20 and November 10, equivalent to the average expansion velocity of the Hay GWS although the initial rate of expansion of the 1933 Great White Spot was considerably greater. Afterwards, we see a surprising decrease in size. It is probable that this observation represents nothing more than the nucleus of the spot, although there is nothing in the observations which allows this conclusion to be checked.

TABLE 2

The size of the Torrell spot calculated from visual meridian pass observations.

Date	*Size (km)*	*Observer*
20-10-90	15,000	Daniel Verde
04-11-90	26,000	Josep-Maria Gómez
10-11-90	49,000	Víctor González
22-11-90	17,000	Víctor González, Francisco Hernández, Pablo Rodríguez

The evolution of the form of the GWS and the disk of the planet

There are very few details available of the initial form of the GWS. The first image or drawing which seems to have been my own independent discovery observation made on September 29.86, 4.7 days after Wilber's discovery. Wilber himself has commented that, between September 25 and 28, the spot became significantly more diffuse and lost brightness; this observation may be considered doubtful though because the brightness of the spot depended very strongly on its position on the disk. When on the meridian it was very bright indeed, but the brightness diminished rapidly more than 40° from the meridian and the spot always disappeared well before reaching the limb. My observation showed what was later seen to be the start of the rapid evolution of the shape of the spot, so evident both in drawings and in photographs made over the next two weeks. The spot had the form of an ellipse twice as long as wide and occupied two-thirds of the width of the EZ (Figure 10); a sketch made by Eva Redondo-Terrón a few minutes earlier when confirming the appearance of the spot confirms this size and shape. The trailing edge was well defined, although the leading edge was fairly nebulous and, at times, a westward extension could be seen. In later observations, this extension became ever more pronounced. What was evidently happening was that large quantities of material were outflowing from the spot in the direction of the rotation of the planet. What started on September 29 as a fuzziness of the preceding edge of the spot had become a large bulge by October 2 and a complex of loops and whorls by October 5.

In a photograph taken at 21.13 UT (immediately after the first visual observations, but before the sketch shown in Figure 10 was made) the GWS is just visible, slightly past the meridian. Wilber reported that the spot was positioned on the southern edge of the NEB, without further comment. On September 29, there was clear activity in this belt: the GWS intruded some considerable distance into the NEB, although various observers reported that at the same time the NEB had extended itself towards the south. To the east the belt formed a plateau which wrapped round the trailing edge of the spot. This structure was reported on September 29 by Nicolás González. On October 1, Blanca García commented that: 'the Equatorial Belt was to a great degree consumed by the white spot'. An important datum recorded on this date is the fact that the GWS was not visible until it reached just 43° from the meridian, which implies that, contrary to expectations, it was at a deep level in the atmosphere. I had also observed this effect on September 29, when, to my considerable surprise, the GWS disappeared approximately 38° from the meridian; had the spot not been confirmed by three other people I think I might well have thought that my observation of it was just the result of a feverish imagination.

This aspect of the Great White Spot's behaviour surprised various people initially. When reporting the 'discovery' to Alan Heath on the evening when the observations were made, I recall him asking if the spot was seen to project away from the limb. Intuitively, it seemed likely that this would be a phenomenon of the upper atmosphere of the planet. This idea was to some degree

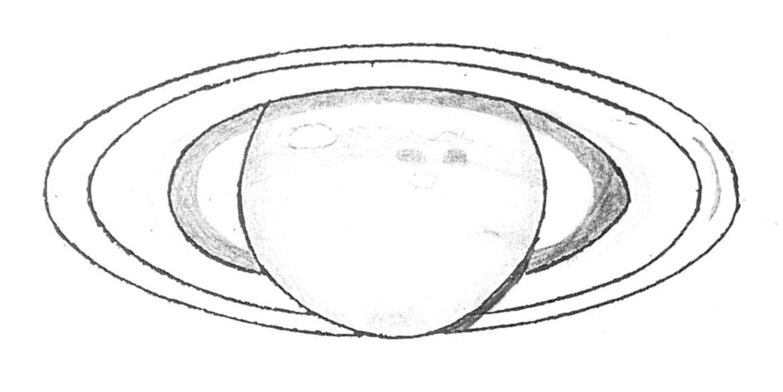

Figure 10. A sketch of the original appearance of the GWS as observed by the writer on September 29, with a 51-cm reflector, ×270. This appears to have been the first observation of the Wilber GWS made outside the United States and constituted an independent discovery of the spot.

reinforced by Augustín Sánchez-Lavega's models of humid convection in the Saturnian atmosphere. Sánchez-Lavega modelled the Great White Spots of Saturn as huge convective outpourings from the deep levels of the atmosphere. However, the fact that the spot disappeared so close to the meridian demonstrated clearly that the spot could not be above the upper level of haze in the atmosphere of the planet.

All the observers who observed the Wilber spot on October 1 noted the considerable change in the North Equatorial Belt, it being much broader and darker in front of the GWS than behind it, where instead, it was much fainter and narrower. These observations are supported by professional CCD imaging which shows how the outflowing material started to submerge the NEB.

The spectacular drawing made by Jean Dragesco on October 2.82 shows the rapid evolution of the GWS (Figure 11). The nebulous form of the leading edge is seen as a narrower extension of the main body of the spot. The GWS now occupied the entire width of the EZ and intruded to a large degree in the NEB. On the other hand

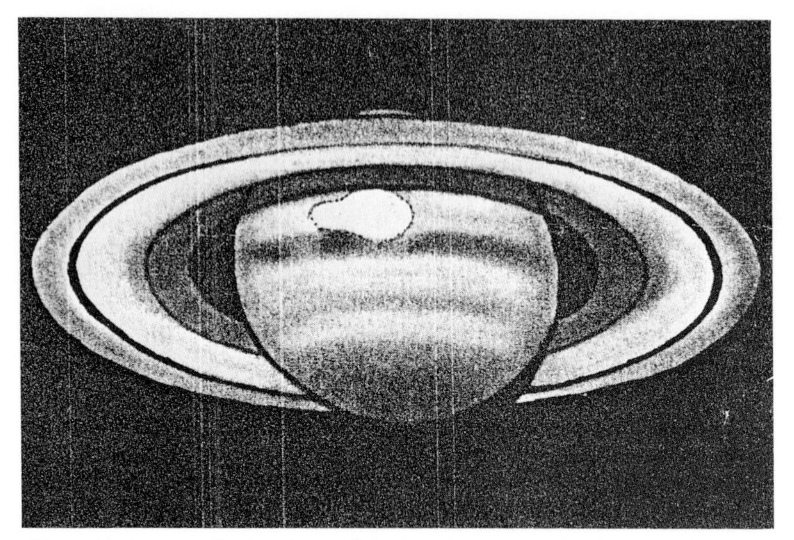

Figure 11. A spectacular drawing made by Jean Dragesco on October 2. The fuzziness of the leading edge of the spot seen in Figure 10 has now developed into a considerable elongation.

though, there is no longer a marked asymmetry in the width and darkness of the NEB.

The observations made by Jordi Aloy from Barcelona on October 5.78 (Figure 12) followed the trend of constant changes. On the one hand there was a doubling of the NEB, probably due to material from the spot covering part of the belt similar to the way that, during its famous disappearance in 1989, the South Equatorial Belt of Jupiter was split into two very fine belts, and bright, high cloud completely obscured the centre of the belt. Whilst, on the other hand, the development of the form of the GWS was now even more extreme. The extension to the leading edge had grown considerably, while the trailing edge remained well defined. We can, in fact, see that the main body of the spot (the large oval) had hardly grown at all; all the development was in the sense of the rotation of the planet, in which we see the development of a pronounced neck, giving the spot the appearance of an exaggerated pear. On the same night, Francisco Campos confirmed this general form.

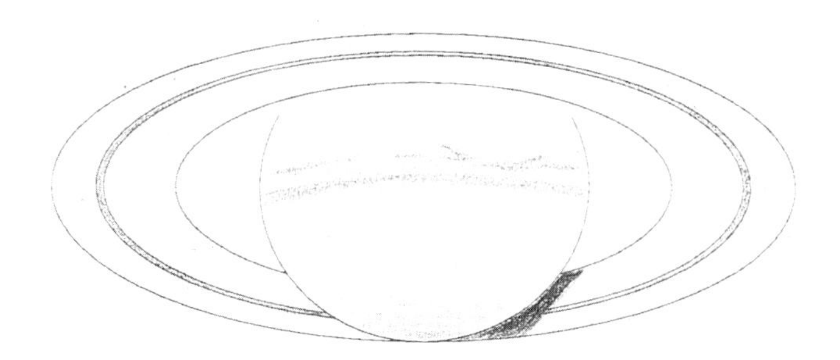

Figure 12. A drawing made by Jordi Aloy on October 5 using a 20.5-cm reflector, ×240. The rapid evolution of the spot is now very clear.

The spot was extremely easy to see, in part due to its expansion. On October 7, Oswaldo González was able to see it clearly enough to make an accurate meridian timing with a 6-cm refractor from Gran Canaria. Despite the fact that Oswaldo has phenomenal eyesight for fine planetary detail, this shows just what a conspicuous object the Great White Spot had become.

On October 10, the disk showed some spectacular changes (Figures 13 and 14). The original GWS now occupied the full width of the disk, such that a meridian transit was not observable. Instead, three new nuclei of activity appeared, superimposed on the Wilber GWS. The first of them (A), was initially detected by myself with the 51-cm Mons Telescope of El Observatorio del Teide, and confirmed very soon afterwards by Oswaldo González and Víctor González; it crossed the meridian at 21.20 UT. This nucleus was small, occupying the third of the width of the EZ and very brilliant indeed, being clearly visible superimposed on the still very bright background of the original GWS. At 21.45 UT, Oswaldo González and Víctor González (observing with the Mons Telescope) and Blanca García and Pablo Rodríguez (with the 40-cm Vakuum Newton Teleskop, also in El Observatorio del Teide) simultaneously detected a second small and extremely brilliant spot (B). This nucleus crossed the meridian at 22.00±2 minutes. When comparing our notes afterwards, Víctor, Oswaldo, and myself realized that we had all independently noted the presence of a third spot (C), this time larger, not quite so bright and occupying the full width of the EZ. This third nucleus crossed the meridian at 22.25

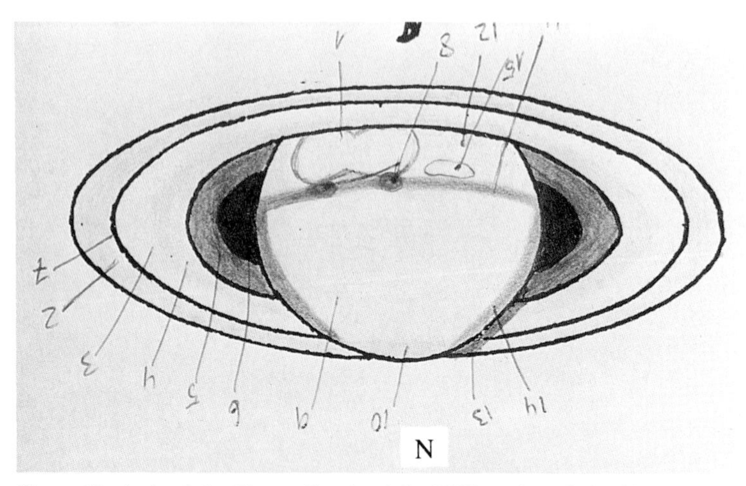

Figure 13. A sketch by Blanca García of the GWS made with the 30-cm Vacuum Newton Telescope of Teide Observatory, Tenerife, on the evening of October 10, 1990. This drawing shows the formation of two nuclei within the original GWS, one of which was very small and brilliant and is seen again in the following figure.

UT approximately. The GWSf was close to, but clearly separated from, the third nucleus. Despite this, a fine line of bright material could be seen, attached to the edge of the NEB, which extended from the trailing edge of the Wilber spot for at least thirty degrees more. This line of material seems to have spread behind the Wilber spot, expanding out like a tidal wave along the belt. The NEB itself was highly active, with several condensations and projections into the EZ. Both nuclei, 'B' and 'C', were associated with different projections, each being in contact with the trailing edge of one. One projection, which separated nuclei B and C, was clearly visible in a CCD image taken in England by Terry Platt at 18.11 UT on October 11. Nuclei 'A' and 'B' were sufficiently bright to be clearly visible to Orlando Rodríguez with a 60-mm refractor on October 10 and 11 respectively.

The sequence of observations made on October 13 by Pablo Rodríguez, Javier Sánchez, Natalie Gúillen, Oswaldo González and Daniel Verde demonstrate the continued evolution of this activity. Of the three nuclei. 'A' remained small and condensed, 'B' was seen by Oswaldo as the nucleus of a bulge in the extended original GWS (although other observers saw nothing more than the

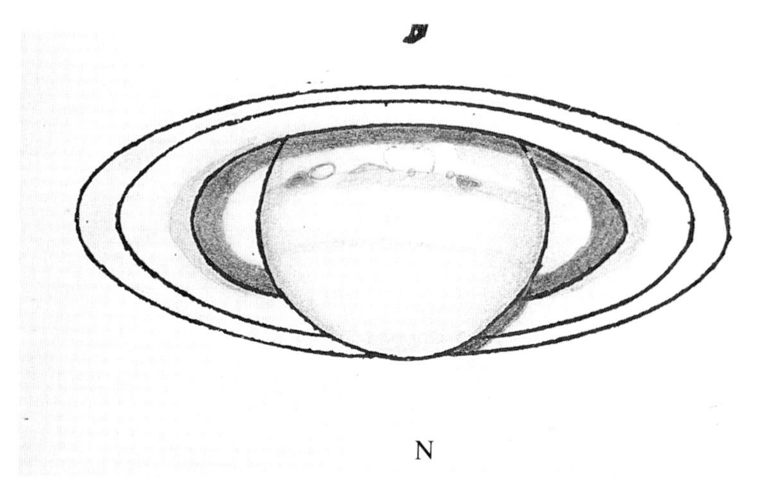

Figure 14. A drawing of the GWS by Mark Kidger made about 40 minutes after the observation shown in Figure 13. Two small, brilliant spots can be seen as well as the trailing edge of the original GWS. This observation was made with a 51-cm reflector, ×270.

bulge). Nobody recorded nucleus 'C' as a spot, although Daniel Verde suspected that he could see it and both Pablo Rodríguez and Oswaldo González saw another bulge in the Wilber spot which corresponds to the position of this nucleus. It appears that, without doubt, nucleus 'C' was now dispersing itself in the mass of the original GWS. No observer saw the GWSf, which must have been in a longitude greater than 060°. Oswaldo González and Daniel Verde (Figure 15) clearly saw a narrowing of the material forming the Wilber GWS at longitudes greater than 330°; the material followed the border of the EZ with the NEB in a form similar to a wave. All the observers saw the NEB much narrower and fainter than usual. The increase in size of spot 'B' was confirmed by Orlando Rodríguez on October 16, on which night he made a series of observations of the approach of this spot to the limb.

Orlando Rodríguez's observations confirm that the brightness of Great White Spot diminished rapidly as it left the centre of the disk. The intensity estimates were: −2.5 at 22.10 (37° from the meridian), −2 at 22.20 (43° from the meridian) and −0.5 at 22.50 (61° from the meridian). These values are combined in Figure 16 with an

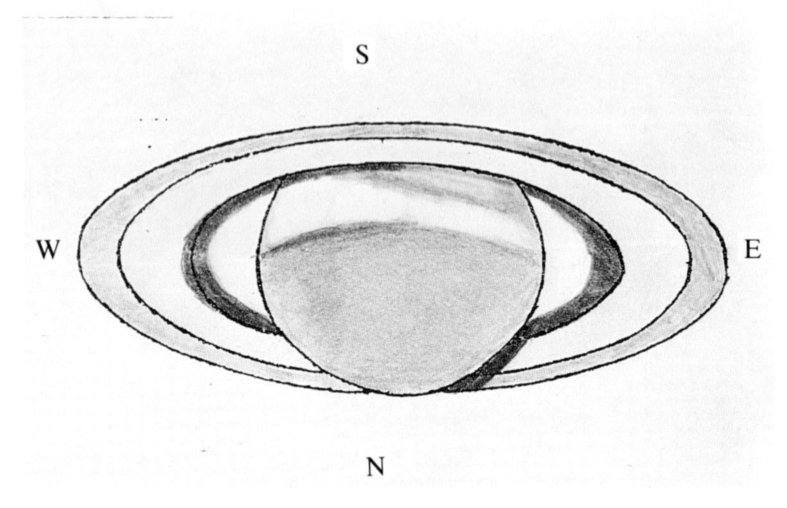

Figure 15. A drawing made by Daniel Verde with the 51-cm reflector on October 13, 1990. The longitude of the central meridian is slightly greater than that of Figure 14, but both figures show the trailing edge of the original spot. The change in the form of the spot in three days and the narrowing of the equatorial belt are very obvious.

intensity estimate which Daniel Verde made of the spot on the meridian to show how the spot faded rapidly even at comparatively small distances from the meridian. However, we see that the spot was apparently rising up within the atmosphere. On September 29 it had vanished at 38° after passing the meridian; on October 1 it appeared at 43° before the meridian crossing and, on October 16, nucleus 'B' was clearly visible, although fading rapidly, as far as 61° from the meridian. These observations imply that the amount of haze above the spot decreased by a factor of 1.6 and hence that the spot had risen a considerable amount in the atmosphere of the planet.

It appears that new activity started on October 20. Danielle Dalloz and Daniel Verde noted the presence of an ellipse in longitude 70° approximately. As it was visible in an 8-cm refractor, there is no doubt that it was an important spot. Danielle Dalloz also saw this feature on October 26. Daniel Verde also observed a prominent nucleus which crossed the meridian at about 22.00 UT

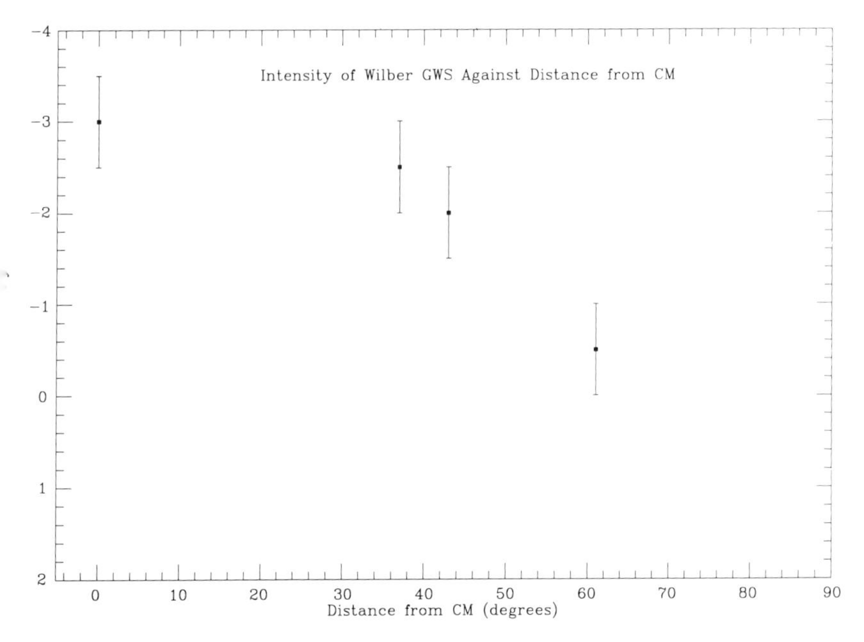

Figure 16. Variation of the intensity of nucleus 'B' with distance from the meridian from estimates taken on October 16. The fading is very pronounced, but less so than immediately after the formation of the GWS.

(Figure 17). The longitude of this feature was 131°, in agreement with the longitude of a major outburst discovered on November 2 by Sebastìa Torrell in Barcelona and confirmed, within minutes, by Josep Ma. Gómez. The position and rotation period of the spot show that the nucleus seen by Daniel Verde was a prediscovery observation of the Torrell spot: Daniel however, in one of those accidents that happen from time to time, but are maddening when they do, did not realize that his observation was significant and did not report his observation until some two weeks later. (Daniel was somewhat mortified by the news that he had missed discovering this spot by failing to recognize the significance of his observation. In 1990 he had also picked up the return of the South Equatorial Belt of Jupiter before this was reported officially on an IAU Circular but, again, had not reported his observation in time to have a chance of claiming credit for the discovery. Like various other

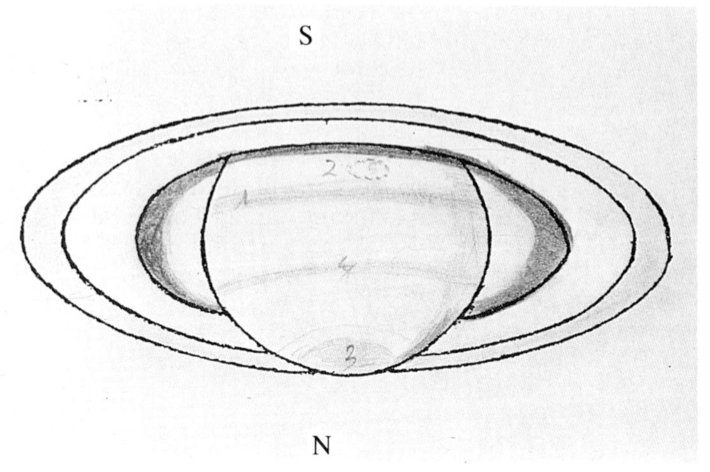

Figure 17. This drawing made by Daniel Verde through a 20-cm reflector on October 20 shows the spot later discovered by Sebastìa Torrell. This observation shows the importance of reporting any anomaly as quickly as possible.

amateur visual observers in the last few years, he has learnt the importance of prompt reporting of observations the hard way. Despite this, it is to his credit as a young, talented and very enthusiastic observer, that he managed to pick up two important phenomena in this way in a matter of three months.)

On October 27, Daniel Verde commented that the EZ appeared to be recovering its normal appearance and darkening significantly. The last drawings which I have seen were made by Daniel Verde on November 12 and 13; on the 12th he did not note any unusual activity although the NEB was narrow and faint; on the 13th he saw the Torrell spot (although not without difficulty) as a large diffuse structure. After this date, a combination of poor weather and limited visibility of the planet at the start of the night have caused grave difficulties when trying to make disk drawings, although there are some later meridian timings.

Post-conjunction observations in 1991

Previous experience of Great White Spots of Saturn has demonstrated that the activity may last for a very long time after the original eruption. In 1906 the disk remained active for some two years. It was thus felt to be very important to continue observations through 1991 and, probably, 1992 to get a complete history of the 1990 Wilber Great White Spot. The observations made through 1991 did, as expected, show that the activity continued for more than a year after Wilber's original observation.

Post-conjunction observations were made by a number of Agrupacíon Astronómica de Tenerife members early in March 1991. One feature similar to the pre-conjunction Great White Spots was observed in the EZ by the writer on March 9 and 18 at longitudes 082° and 093° respectively. On the former date it was small and very brilliant. By March 18 it was now much larger and more diffuse. Despite various alerts to other groups, no other known observations were made of this feature, although Alan Heath believed that more observations would come to light when observers handed in their observations for the year. Another striking aspect of the disk was the huge amount of fine structure in the Equatorial Zone; the whole of the EZ looked 'grainy' as if made up of a fine honey-comb of individual cells.

One striking aspect of the disk though, throughout the spring of 1991, was the fact that contrast was minimal. Even the North Equatorial Belt seemed to be very diffuse and hazy. The appearance of the disk was that of one veiled by an intense, high altitude haze which extended over the entire visible hemisphere. On previous occasions, an equatorial Great White Spot has been preceded by a global disturbance. This was no exception. Over the course of spring of 1991 it became obvious that the NEB was becoming very much broader than in 1990 and seemed to be displaced towards the south. The Equatorial Zone was bright, as in 1989, but very much less so than it had been as the GWS spread around the disk of the planet.

The developing activity on the disk culminated in the summer when various groups reported that the North Equatorial Belt had split into two components. On August 31, 1991 a group consisting of Jorge González, Rafael Barrera, Alberto Darias, and myself noted that the NEB had split into two well-defined components. It seems that other observers may have noted this previously, although no observation reached me before making a report to Brian Marsden, and Jorge González had observed Saturn with the same telescope and in similar conditions some four weeks earlier without seeing any marked splitting of the NEB. Daniel Verde picked this up independently on September 1 and had also not previously noted any splitting; at the very least it seems that, even if the belt were split previously, the splitting became very much more prominent around this date. This splitting of the NEB follows the sequence of activity noted on previous occasions. Various white ovals and structures were noted in the Equatorial Zone, although it is clear that there were no white spots on the scale of, or as prominent as, the Great White Spot of the previous year. On September 16 I noted the presence of the Equatorial Band for the first time. Other observers such as David Grey and Josep Maria Gómez who looked at Saturn around this date also noted that this normally elusive feature was particularly prominent. Josep Maria Gómez, in particular, noted that there was an extraordinary amount of structure in the Equatorial Zone.

Other observations show that this may have coincided with a significant change in the colour of the North Equatorial Belt. Through 1989 and 1990, its colour, established by colour filter observations, had been neutral. In October and November 1991 though, colour filter observations which Daniel Verde has passed me showed that the belt had now got a strongly ruddy tint, showing a marked colour change.

It seems that, by conjunction in 1991, the activity on the disk of the planet was dying out rapidly. Whilst we may expect to see some residual activity in early 1992, it is likely that the Great White Spot and its effects as such, has probably disappeared at last. We will probably have to wait until early 2017 to get another sight of a Great White Spot on Saturn.

CCD imaging from La Palma

These observations were made on three nights over the period October 21–26 with the 2.56-m Nordic Optical Telescope (NOT), an f/11 Cassegrain reflector equipped with a CCD camera with a 520 × 520 Tektronix CCD chip, giving a 0.20 arcsecond pixels size. All exposures were of 0.1 seconds duration, using B, V, R, I, and Z filters. The rather poor seeing reported for some of the images was due to the fact that the observations were made just after opening the telescope, with temperature equilibrium not yet established. On October 22 and 26 the seeing was 0.6″ for long (5-minute) exposures. On October 24 it was 1.1″ for long exposures. The observations reported here were made in early evening twilight before starting the scheduled programme of observations of BL Lac objects with the telescope, in time granted by the Nordic countries'

Figure 18. Observing the Great White Spot with the 2.56-m Nordic Optical Telescope in La Palma. This photograph, taken by Leo Takalo on October 26, shows the writer looking at one of the CCD images of Saturn coming up on the TV screen.

time allocation panel. All the observations were taken by Mark Kidger and Dr Leo Takalo of Turku University Observatory, Finland.

October 22, LCM (System I) 280°

In 'B' (blue) the NEB was almost invisible; it was only seen as a type of contrast effect between the EZ and the NTrZ. The border with the EZ showed some degree of structure. There was a brilliant narrow band between the western limb and the meridian. This was attached to the northern edge of the NEB and filled the space between the NEB and the NTB (which was barely visible). The NPR was visible, although not very prominent. The EZ was full of structure, a large bright oval within it being the most prominent structure. There appeared to be a bright spot on the border with the NEB, some thirty degrees east of the meridian. In 'V' (green) the NEB was very narrow and full of structure, having some prominent projections into the NTrZ. There was a narrow, very brilliant band to the east of the disk, in the NTrZ. There was also a dark and very elongated oval in the EZ, almost on the meridian, which was about a third as wide as the EZ. The centre of the EZ was much brighter than the limb regions. The NPR was visible, but not prominent. In 'R' (red), the NEB was very narrow and darker to the east of the disk; to the west it was almost invisible. The NPR was dark and surrounded by a narrow, bright, but not very contrasted band. The EZ was split in two by a dark band, visible in the centre of the disk and almost exactly in the centre of the EZ. It is possible that there were several bright spots between this band and the NEB. In 'I' (infrared) the NEB was rather broader than in other colours, although it still contained a lot of structure. The same band/oval was visible in the middle of the EZ. Between it and the NEB two bright nuclei were seen. The NPE was very dark and surrounded by a prominent bright band. In 'Z' (1 micron), both the NTB and the NNTB were faintly visible. The NPR was very dark. The NEB was much broader than in the blue bands and much darker to the east. A dark oval was visible in the EZ, which had a bright spot at its western end and which crossed the entire disk apart from the zone of the white spot.

October 24, LCM (System I) 165°

In 'B',the most prominent structure was a bright band superimposed on the NEB, although the NEB itself was invisible. The brightest part of this band was to the west of the meridian. The EZ was much less bright and seemed to have a faint dark band in the middle. The NPR was visible although hardly prominent. In 'V' the disk appeared almost as in 'B', although two brighter nuclei could also be seen in the band superimposed on the NEB. The NPR was less dark than in 'B'. In 'I', the bright band was resolved into a string of brilliant knots; what was apparently the NEB could be seen to the north of this band. The EZ was split in two by a dark band which crossed almost the entire disk. The NTB was just visible, although the NPR was very dark and surrounded by a brighter ring. In 'Z', the disk was almost identical to that in 'I'; the only differences being the increase in darkness of the NPR and the decrease in the contrast of the ring surrounding it, and the fact that the NEB was significantly broader than in 'I'.

October 26, LCM (System I) 055°

In 'B', the EZ was now identical in brightness to the NTrZ. The NEB was still hidden below a bright band, which had a strongly wavy form, both on the border with the EZ and with the NTrZ. There was a darker line (the NEB) visible on the southern edge of the bright band. The NPR was visible, but not prominent. In 'V', the disk was very similar to that in 'B' although, on this occasion, the dark edge to the bright band was to the north. Again, this band was strongly waved. The EZ was significantly less bright than the NTrZ. In 'R', the bright band covering the NEB only extended from the western limb to a point some twenty degrees past the meridian, but was full of fine structure. The NEB was visible to the north of this band, but very faint. The EZ was slightly brighter than the NTrZ, whilst the NPR was dark and surrounded by a slightly brighter ring. In 'I', the bright band extended very little past the meridian. To the west, the NEB had its normal width but, to the east, was only visible to each side of the bright band. This band was very brilliant and split into many individual knots. The EZ was slightly brighter than the NTrZ, whilst the NPR was very dark and surrounded by a slightly contrasted bright ring. In 'Z', the NEB was seen split in two and the brilliant band extended from the meridian to the western limb and was full of structure and waves. The EZ was much brighter than the NTrZ, whilst the NTB was broad and very diffuse. In contrast, the NNTB was narrower and darker. The NPR was very dark and surrounded by a band slightly brighter than the rest of the NTZ.

The Nature of the Great White Spot

Up to now I have said little about the physical nature of the Great White Spot. When the press picked up the story it was reported as a great storm of ammonia in the upper atmosphere of the planet. Its cause though is not at all well understood. Augustín Sánchez-Lavega has estimated that a typical Great White Spot forms initially at a level in which the atmospheric pressure is around 1 Bar (i.e.: 1 atmosphere) at a temperature of some 125 K, below the layers of tropospheric and stratospheric aerosol. Over the course of its development it may rise nearly to the top of the troposphere, increasing its altitude in the atmosphere by some 70 or 80 km. As it does so the temperature of the atmospheric layer in which it is resident cools with the considerable fall in pressure until at its maximum altitude its temperature will be about 80–90 K.

The height at which the Great White Spot forms corresponds to that of the cap of ammonia in the atmosphere, the highest of the three distinct layers which are to be expected; it is this that gives rise to the 'ammonia storm' nickname applied to the 1990 spot. The

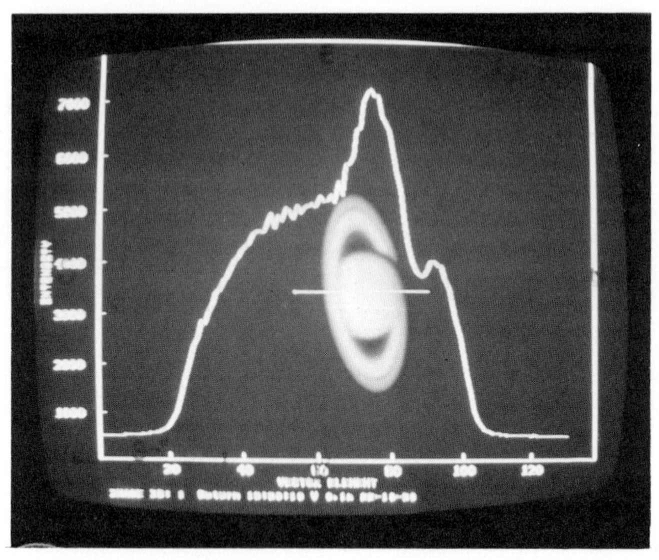

Figure 19. A CCD image of the GWS taken on October 26 with the Nordic Optical Telescope. A scan has been taken through it on a north–south axis showing how the brightness varies across the disk (photograph by the writer).

Figure 20. The CCD image shown in Figure 18. This image is a 0.1 second exposure in the Gunn-Z filter. The seeing was about 0.6 arcseconds. A considerable amount of fine structure can be seen in the spot.

mechanism for the formation of the spot is thought to be a prolonged period of insolation (that is, exposure to direct sunlight) of the northern hemisphere of the planet. This period of prolonged heating gives rise to a convective instability in the atmosphere. A deep convective instability will sweep deep layers of ammonia up from the lower troposphere to the higher layers. The exact mechanisms for this though are unclear. For example, the question remains why this should occur in the form of a localized spot rather than as a global disturbance, even though the Great White Spots turn into global disturbances over the course of months or years.

The coincidence of the outbreaks of Great White Spots with the summer solstice in the northern hemisphere of the planet and the apparent periodicity close to the 29.46-year Saturnian orbital period have been widely commented upon and are seen as proof that the Great White Spots are related to the incident solar radiation. However, the situation is not as simple as it seems. Previous Great White Spots have appeared in latitudes which were receiving the maximum quantity of incident solar energy however, there is no correlation between the total amount of solar heating of the atmosphere of Saturn when a spot appears and its latitude in the atmosphere. The 1876 and 1903 Great White Spots occurred at times when the total amount of incident solar energy on the upper atmosphere was the same but then appeared at completely different latitudes within the atmosphere: the 1876 spot at 8° north and that of 1903 at 33° north of the equator. The 1933 spot appeared at almost the same latitude as that of 1876 (5° north), but at a date when the total incident energy was somewhat less than in 1876. Only the 1960 spot seems to be clearly separated in that it appeared at a very high latitude (+58°) when the amount of incident solar energy was 25 per cent greater than in either 1876 or 1903.

A further detail which shows that the correlation between the amount of incident solar heating and the formation of the Great White Spots is not as simple as it seems is given by the 1990 GWS. The previous four Great White Spots had occurred between 13.6 and 15.7 years after the deep minimum in the solar heating (i.e. mid-winter). For the 1990 Wilber Spot this interval was the shortest on record, being just 11.9 years, yet the latitude of formation (+12°) was rather more northerly than the two previous equatorial Great White Spots. Up to the 1990 spot the interval between mid-winter and the formation of the spot was 14.65±1.2 years. Whilst the Wilber spot is only 2.4 sigma (2.4 times the standard deviation) from the trend established by the previous four spots, the deviation from the previously established trend is surprisingly large. The four previous Great White Spots appeared, on average, 14.65 years after mid-winter in the northern hemisphere; this is so close to being exactly half a Saturnian year (14.72 years) that it can hardly be coincidence. Perhaps the key factor in the appearance of Great White Spots is not, in fact, that they occur approximately at mid-summer, but rather that they occur almost exactly half a year after mid-winter in the northern hemisphere. However, as has been pointed out, exactly half of the time our view of the northern hemisphere is blocked by the rings. Thus, were Great White Spots to appear in the northern hemisphere at this time also, we cannot see them anyway.

Probably the 1990 Wilber Great White Spot will go a long way to improving our understanding of this phenomenon. At present though it is far too early to hope for scientists to unravel all the mysteries of the Great White Spots of Saturn. Certainly though, for the first time ever we have been able to study a Great White Spot with a variety of different instruments which would have astonished Asaph Hall and Edward Emerson Barnard, the discoverers of the first two known Great White Spots on the planet.

Conclusions

Many aspects of the Wilber GWS have been unknown in previous GWSs, although we should remember that our database on the GWS phenomenon is very poor. The Wilber GWS appears to have evolved much more rapidly than any previous GWS. Its rate of increase in length of 14 per cent per day was more than three times greater than that of the Hay GWS, the only previous GWS to have shown an indisputable evolution of size. The Wilber GWS has also shown at least one, and probably two sudden changes in period, the first, very briefly around October 1 (and perhaps not beyond doubt), the second, after about October 5 or 6, is absolutely clear in the data and implies a sudden reduction of about four minutes in its rotation period. Such changes in period can be interpreted in terms of a strongly stratified atmosphere, with strong gradients in wind speed between different levels. There is evidence that the spot developed at a rather low level in the atmosphere, but rose up through the atmosphere over the course of about three weeks.

We have also seen that the number of different eruptions has been exceptional. Only the Botham GWS in 1960 has shown a comparable number of nuclei of activity. Given that the activity of GWSs in Saturn can last for six months or more, it was hoped that observations of the planet taken after conjunction in February or March 1991 would show further Great White Spot activity; although this did not happen, there was some evidence of small bright spots during the spring of 1991. It is not at all impossible that further eruptions may have occurred during the period of invisibility. The only way of following such activity would have been through spacebound observations (with the Hubble Space Telescope), or visual observations made in broad daylight, with the planet close to the Sun in the sky, but it seems that, unfortunately, no such observations were made.

Acknowledgements

I am no expert in Solar System studies and have thus had to rely on the help of many people too numerous to mention individually. My grateful thanks to all who helped me in presenting this work, particularly Josep Maria Gómez, Augustín Sánchez-Lavega, the A.A. de Tenerife and the coordinators of various other groups. It is a pleasure to acknowledge the help of the Instituto de Astrofisica de Canarias through their concession of generous amounts of time with the 51-cm Mons Telescope in Observatorio del Teide, and for the granting of observing time made by the Spanish Comité de Asignaciôn de Tiempos (CAT) to the A.A. de Tenerife.

Light Pollution, Radio Noise, and Space Debris

A Hazard for Observational Astronomy

D. McNALLY

In order to understand the many parts of the Universe, it is essential to *observe* astronomical objects, whether they be asteroids in the Solar System or the most distant galaxy in the Universe. The word 'observe' implies a severe logistical constraint on the astronomer. Astronomers cannot experiment with the objects they study. Astronomers are therefore denied one of the biggest advantages of the laboratory – the ability to change measured parameters in a controlled way. Astronomers must do the best they can by analysing such signals – normally electromagnetic radiation (light, radio waves, etc.) – as are emitted by the object under study. It is rather like deducing the internal structure of an orange simply by looking at one.

The terrestrial astronomer must observe through the Earth's atmosphere. Apart from breathing, the atmosphere has little to offer the astronomer. It is in a constant state of change, it is frequently cloudy. The unsteadiness of the atmosphere shows to the naked eye as stellar twinkling. This is the phenomenon of seeing – the continual change in the refractive index of the atmosphere reflects in small changes of position of the essentially point source of light which is a star. At many urban sites a stellar image can have an apparent diameter of 5″ to 10″, because of seeing; but in the rare, superb, observatory sites apparent diameters can sometimes get as low as 0″.2. Starlight is also absorbed in the Earth's atmosphere because it has a finite gas density and has a permanent dust population. Radio astronomers have somewhat different problems with the atmosphere which are just as tedious. When you think about it, a rational being would not undertake the science of astronomy from the surface of the Earth, given the nature of its atmosphere. Astronomers, by being enthusiasts with insatiable curiosity, have done wonders in beating the constraint posed by the atmosphere.

But mankind is putting ever more obstacles in the way of being able to observe faint astronomical objects from the surface of the

Earth. Figure 1 is a picture of 'The Earth at Night' showing the effects of many forms of outdoor lighting. The picture was put together by W. T. Sullivan III using pictures of the night-time Earth taken by US Meteorological Satellites. We have outdoor lighting at night in order to let us see what we are doing – yet Figure 1 shows that a great deal of that light escapes uselessly upward. That escaping light on its passage into space is scattered by the dust and dirt in the atmosphere brightening up the sky and so making observation of the faintest celestial objects impossible.

The scattering of the light by dust and dirt would occur naturally but industrial processes have compounded the problem by the amount of filth that such operations create – dust from quarrying,

Figure 1. 'The Earth at Night' prepared by W. T. Sullivan III. (Reproduced by kind permission of the International Dark-Sky Association and the Astronomical Society of the Pacific.)

from factory chimneys, from building sites, from metal working – the list is endless. Fortunately, such dust does wash out of the atmosphere (Did I write 'fortunately' when you consider acid rain?) and, as yet, it seems industrial effluent has not had a significant impact at the remotest observatory sites. Nevertheless it must continue to be a major concern.

The upward escape of outdoor lighting need not occur. We can have well lit streets and reasonably dark skies – skies, that the serious amateur astronomers can utilize for their observing projects, that the urban professional observatory can still undertake front-line research, and that the non-astronomer can enjoy for the spectacle of the Milky Way and a starry sky. Have you counted the number of stars you can see with the naked eye from the centre of any large city? You will not exhaust your fingers! Yes, rather less than ten.

Good outdoor lighting can place light where it is needed – on the road and on the pavement and *not* in the sky. With at least 30 per cent of all street lighting going upwards, if that light were reflected downwards we could save power and still have the same light levels as we now enjoy. Perhaps 'enjoy' is too strong a word, for much of our outdoor lighting is overkill – there is too much glare and a reduction in light levels to get rid of that glare would in fact enhance the visibility which is what outdoor lighting is all about. So there is something in better lighting for all of us, taxpayer and astronomer alike, plus the advantage of using less power and so putting less effluent from power stations into the atmosphere.

But astronomers can be helped further by using the right type of lamp for outdoor lighting. An ordinary tungsten bulb which we all use in our homes emits white light. White light is made up of light of all colours from red to violet. The spectrum of a tungsten lamp is just like a rainbow with light of all colours. But a low-pressure sodium lamp only emits two very narrow spectrum lines close together in the yellow (and two more weak spectrum lines far in the violet). The rest of the spectrum has no light in it at all, so if astronomers forgo investigation at the wavelengths of the yellow lines of sodium they have a large range of wavelengths to study – fortunately much interesting stellar spectroscopy is in the blue region of the spectrum. Again, everyone has something to gain. As a bonus, our eyes are most sensitive to green/yellow light. Clearly, there are situations when sodium lamps are inappropriate – for example, in situations where colour balance is important – a small amount of white light added to an overall sodium illumination can restore the colour balance.

Only 50 per cent of all escaping light comes from street light. The rest comes from illumination of signs, illumination of sports stadia, playing fields, tennis courts, vehicle headlights, and the escape of domestic lighting through uncurtained windows. Proper lighting design can also minimize the upward escape of all such lighting and anyone can pull their curtains at night (a move which also helps to cut down on heat loss from the central heating).

The situation facing the radio astronomers is not as good as the bad situation facing the optical astronomers. Again, mankind's activities produce a great deal of radio noise – TV sets, refrigerators, ovens, milking machines, internal combustion engines, electric trains, etc., etc. The radio din is extensive. The map of radio noise would be very similar to the outdoor lighting map. But the radio astronomers face an even more threatening situation. Although special wavelength bands are reserved for the exclusive use of radio astronomy, modern communication – mobile phones, radio communications, high definition TV, satellite transmission – is putting increasingly severe pressure on the available band space. The great advances in radio astronomical techniques in the last decade mean that radio astronomers are feeling the limitations of their rather meagre allocation of exclusive band width.

Radio telescopes, unlike their optical counterparts, can pick up signals outside the immediate line of sight to a celestial object – the side-lobe responses. A terrestrial or satellite transmitter may be picked up by such a side-lobe. Although the side-lobes have lesser sensitivity, the signal strength received from a transmitter may be many orders of magnitude stronger than the desired celestial signal. Again a transmitter may be operating at its allocated frequency but may also generate harmonics in the reserved radio astronomy wavelength bands. The transmitted wavelength may drift and cross into a reserved wavelength band. Today's commercial pressures are such that, like land through the ages, ownership of bandwidth can mean wealth.

Radio astronomy is represented on the International Telecommunications Union (ITU) and receives considerable support from the ITU. Nevertheless with a growing range of interstellar molecules to be observed, with the ability to measure cosmological red shifts, with the increased precision of positional measurement now available to radio astronomy, to name but three, radio astronomy feels the confines of its reserved bandwidth and could utilize much more. Yet the commercial pressures are attempting to erode the little bandwidth radio astronomy has got.

So when you next use your mobile phone to ask for your dinner to be kept warm from your long motorway tail-back and then use your radio garage door control when you eventually arrive for a welcome dinner warming merrily in the oven, think on the four ways you have been causing radio interference before you switch on the fluorescent light in your uncurtained kitchen.

Of course, the answer to all this is to go into space – or is it? Major sources of radio noise are already in space, such as telecommunications and navigation satellites. The major observatories now in space are spectacular successes but are enormously expensive to build and operate. It is a truism that space astronomy generates an order of magnitude more than ground-based astronomy. There can be only relatively few astronomical observatories in space. But even these are not safe. Since that October day in 1957 when the first Sputnik was put into orbit – over 4000 satellites have been launched (an average of one every three days since Sputnik). About half have fallen back to Earth. Of those that remain most are obsolescent and no longer function. Along with these satellites there are many bits of associated hardware such as spent motors, almost empty fuel tanks, and bits and pieces such as instrument covers. A total of over 7000 objects are routinely tracked, and indeed, the Shuttle Discovery had to take evasive action recently to avoid such a piece of tracked, space hardware. But with so many objects in orbit, collisions must occur. Fuel tanks have blown up with the accidental ignition of remaining fuel. This means that as well as the tracked and known objects there is a great deal of debris which cannot be tracked. Essentially, debris larger than 10 cm can be tracked – smaller sized pieces cannot. A piece of debris as small as 3 gm travelling at 10 km/s^{-1} can inflict as much damage as a crashing car weighing 1 metric tonne travelling at 60 mph. Already a shuttle windscreen has been pitted by collision with a flake of paint. The Hubble Space Telescope is considered to have a 1 per cent chance of a catastrophic collision during its seventeen-year lifetime. It has a much greater chance of more limited damage. Astronomy from space therefore has its own particular hazards.

Space debris also poses problems for the terrestrial astronomer. Satellites and space debris, like any other bodies in space (e.g. the Moon), reflect sunlight. There are now so many satellites and bright debris that a single deep-sky survey plate can accumulate four or five trails in the course of a 2-hour exposure. At least 30 per cent of such deep-sky plates are ruined by such trails. Photometric searches for transient phenomena such as γ-ray bursters can be confused if a satellite or piece of debris passes across the field of view. Although photometric fields are small by comparison with deep-sky plates, crossing by satellites is a common occurrence. Sensitive photometric detectors may be damaged beyond repair if accidentally exposed to sunlight reflected by a bright satellite or piece of debris.

There is a worry that pressure for space art will both increase the number of objects in orbit and lead to a brighter sky. There is no point in putting art in space unless it can be seen. To be seen easily it must appear to be about the same angular size as the Moon and about as bright. For the period of about one week either side of full Moon, no faint object work can be undertaken at observatories. Art is usually planned to have an orbit where the object comes round every two hours. Can you imagine the effect on faint-object astronomy with a sky tastefully decorated with space art? Faint-object astronomy would be effectively at an end. Other bizarre suggestions have been made for the use of space, such as placing the cremated remains of our predecessors into brightly burnished canisters to then be launched in orbit. Certainly late Uncle William would have a striking memorial as his remains flashed across the sky. To astronomy his conveyance would be another piece of light-polluting hazard.

Although the threat to continued astronomical observation of faint celestial objects is real and will continue to grow for the foreseeable future, there are hopeful straws in the wind. Lighting engineers are growing increasingly concerned about bad lighting – glare can be a hazard and poor luminaries can cause disturbance to normal patterns of living. Again, considerable savings both of energy and money can be made if the right amount of light is directed to the right place. Energy savings with concomitant reduction in the atmospheric degradation caused by energy production is a positive benefit in combating greenhouse effect. The problem of radio noise is also being addressed – for example, the noisy GLONASS navigation satellites are to have alterations made by 1994 to filter out the noise they emit which degrades the detection of the interstellar molecule OH. Steps are now being taken to ensure that spent rocket fuel tanks are emptied of residual fuel to minimize the risk of generation of space debris through explosion.

Despite such encouraging signs, constant vigilance needs to be maintained to take all practicable steps to ensure astronomical observation can continue through stringent protection of the best observatory sites, through effective steps to reduce unnecessary electro-magnetic noise and through ensuring that obsolescent space-craft are removed from orbit at the end of their working life. While astronomy may seem a science remote from everyday experience, knowledge of the science of stars and galaxies is essential to understanding the structure and evolution of the Sun and its planetary system. As dwellers on the surface of one member of that planetary system, such knowledge could be crucial.

What are Noctilucent Clouds?

NEIL BONE

Summer at high temperate latitudes can, in some respects, be regarded as an 'off-season' for observational astronomy. From the latitudes of the British Isles, Scandinavia, and Canada, the sky does not become properly dark during the short night. The Sun skirts not far beneath the northern horizon, meaning that observers in these parts of the world are denied a proper view of the rich, deep-sky fields of the southern sky during June and July. Instead of looking southwards, many amateur astronomers – particularly in north-west Europe – have in recent summers been turning their attention to the bright, northerly half of the twilit summer sky, in search of the delicate wisps of *noctilucent clouds*.

It might seem curious, at first glance, for astronomers to actively seek clouds! Noctilucent clouds, however, are uniquely interesting, lying far above the all-too-familiar 'weather' clouds of the troposphere. The troposphere, reaching a maximum altitude of 15 km above the Earth's surface, is the layer within which the bulk of the atmosphere lies. Movements of the air in the troposphere, responding to solar heating and the Earth's rotation, give us our global weather patterns. Above the troposphere is the stratosphere, where the air becomes more rarefied. The stratospheric ozone layer around 50 km altitude, resulting from the action of solar ultraviolet on atmospheric oxygen, is an important screen against the very radiation which produces it for living organisms on the Earth's surface.

Higher still, we come to the mesosphere, the middle-layer of the atmosphere, extending to altitudes of about 85 km. Temperature falls off markedly with height through the mesosphere, reaching a minimum at the *mesopause* at 85 km altitude: it is close to here, above 95 per cent of the atmosphere, that noctilucent clouds form.

The extremely rarefied uppermost fringes of the atmosphere, from 100–1000 km altitude, comprise the thermosphere, within which the aurora occurs and the principal layers of the ionosphere are found. Relative to the mesosphere, temperatures in the thermosphere rise again, though atmospheric particle densities become very low – comparable to those in the interior of a domestic light-bulb.

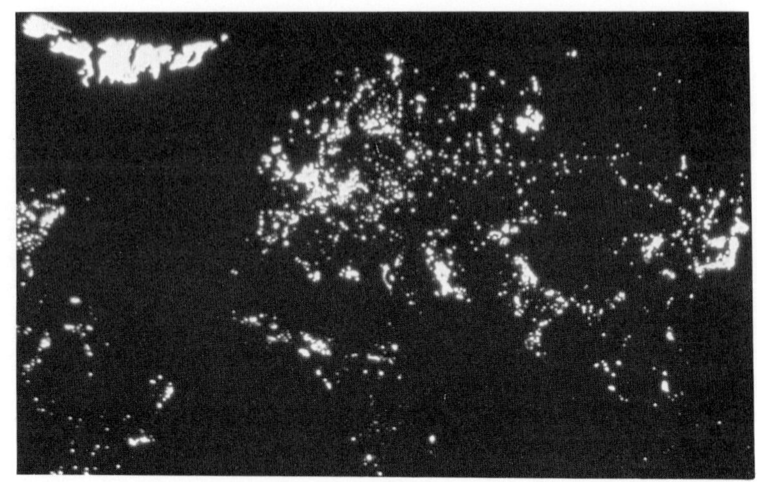

Figure 1. A bright display of Type II noctilucent clouds, photographed from Edinburgh on July 15–16, 1983 by Neil Bone. Exposure was 3 seconds at f/2.8 with a 50-mm lens, on Ektachrome 400 slide film.

The highest tropospheric clouds are cirrus ('mares' tails'), usually found ahead of approaching weather fronts at altitudes of 10–15 km, and comprised of ice crystals. Superficially, noctilucent clouds and cirrus are similar in appearance. Noctilucent clouds, however, occur a great deal higher in the atmosphere. It is by virtue of their great height that noctilucent clouds become visible on summer nights: they are so high that they remain sunlit – literally 'night-shining', as their name suggests – long after even cirrus clouds are immersed in the Earth's shadow (Figure 2).

Direct measurements of the atmosphere by instruments aboard sounding rockets suggest that quantities of water vapour at the mesopause should normally be too low for cloud formation. Noctilucent clouds do, however, form, and may often be readily observed with the naked eye. Admittedly, the sheets of noctilucent clouds are extremely tenuous in comparison with tropospheric clouds – so tenuous, indeed, that they are swamped by the brightness of the daylit sky. Only when they can be seen backlit, by the Sun from just below the observer's horizon, do noctilucent clouds become prominent by contrast against the twilit foreground. The sky illumination conditions which prevail for much of June and July at high latitudes in the Northern Hemisphere favour the observa-

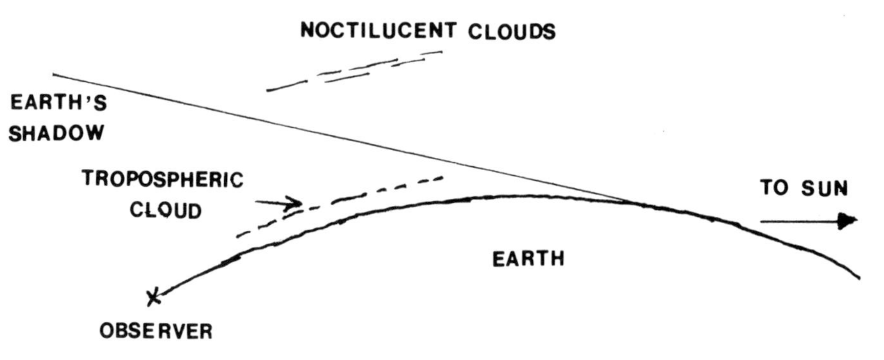

Figure 2. Noctilucent clouds remain in sunlight long after lower, tropospheric ('weather') clouds are immersed in the Earth's shadow. The observer at X, looking towards the bright horizon in the direction of the set Sun, sees noctilucent clouds illuminated against the twilit sky. Note that the curvature of the Earth is not to scale!

tion of noctilucent clouds, whose delicate structures make them a natural target for astrophotographers.

In addition to their æsthetic appeal, noctilucent clouds are of scientific interest, and several professional students of *aeronomy* (a discipline embracing elements of geophysics, meteorology, and astronomy) are actively involved in their investigation. Amateur astronomers, by recording sightings, can contribute results of considerable value to these studies.

What, then, is current understanding of the nature and origin of noctilucent clouds? Most of what is known about noctilucent clouds and their behaviour is based on observations made from the ground, since these clouds form at an altitude too high for balloon studies, and, awkwardly, just a little too low for detailed examination by satellites (though some observations have been made by Skylab astronauts and cosmonauts aboard the Salyut and Mir space-stations). Ground-based observations include triangulation from two or more sites, confirming the great height of noctilucent cloud features, and giving some indication of their movements. It has been suggested, on the basis of such work, that a whole independent 'weather' system, with noctilucent clouds marking out the fronts, exists in the high atmosphere. Noctilucent clouds are often seen to be carried westwards by winds blowing at velocities of 400 km/hr.

Noctilucent clouds appear to be strictly a summer phenomenon.

Not only is this the time when solar illumination conditions most favour their observation, it is also the time when temperatures at the mesopause reach their minimum (around 111 K), as a result of springtime up-welling of cold polar air. This up-welling process is also believed to be responsible for carrying aloft the small traces of water vapour required for noctilucent cloud formation. Noctilucent clouds form principally at high latitudes, then drift towards lower latitudes: it is extremely rare for them to be seen below 50° latitude.

The appearance of noctilucent clouds in thin sheets reflects the limited volume of atmosphere in which they can occur. Cloud formation begins at altitudes of about 85 km, as water vapour condenses around small nuclei. As further condensation occurs, and the particles grow larger and heavier, they fall under gravity. The concentration of noctilucent cloud particles reaches a maximum at 82 km altitude. Further descent takes the particles into a region below the mesopause where temperatures begin to rise slightly, and the condensed water evaporates again.

The nature of the material around which noctilucent cloud particles nucleate remains uncertain to some degree. Obvious candidates would be minute fragments of meteoric debris, suspended in the high atmosphere. The vast majority of meteors become luminous at altitudes of about 90–100 km above the Earth's surface, and the ablation process should leave a substantial amount of suitable debris at just the right altitude for noctilucent cloud condensation. Other possible condensation nuclei include volcanic debris, thrown high into the atmosphere by violent events such as the Krakatoa explosion or the recent eruption of Mount Pinatubo. Suspended in the stratosphere, dust from these events can be a source of spectacular, colourful sunset and twilight phenomena. Indeed, noctilucent clouds were only recognized during the summer of 1885, when many observers were devoting much attention to the twilight spectacles which followed Krakatoa.

A further alternative source of condensation nuclei for noctilucent clouds is provided by the proximity of the ionosphere. Ions leaking over from the E–region could serve as condensation nuclei.

Attempts have been made to capture noctilucent cloud particles for laboratory analysis, using sounding rockets launched from ranges in Sweden and Canada. The Swedish experiments during the early 1960s, using 'Venus Flytrap' rockets, provide a good example of how professional and amateur scientists can work together. While noctilucent clouds form above the launch site at Kronogard, the summer Sun's depression below the horizon is insufficient to allow them to become visible overhead. The researchers were therefore dependent on amateur observers further south, who *could* see whether or not noctilucent clouds were present, to provide the alert that launches might be productive. The 'Venus Flytrap' payload consisted of a sealed chamber, which opened only while the rocket was at noctilucent cloud altitudes. Within the chamber were membranes on to which the particles could adhere, and which were retrieved from the crashed rockets for subsequent detailed examination.

Results of this, and similar work using later rocket launches from Kiruna, Sweden, and from Fort Churchill, Canada in the 1970s, suggested that noctilucent clouds do, indeed, condense around nuclei of meteoric origin, though the other potential sources of nuclei cannot be completely ruled out. The particles are small, probably less than 0.1μm in diameter.

It is difficult to study the size and composition of noctilucent cloud particles using ground-based equipment, thanks to the large amounts of water vapour in the lower atmosphere between the observer and the cloud-layer. Observing from low orbit, thus avoiding this problem, Salyut cosmonauts were able to see noctilucent cloud sheets as thin, edge-on layers when illumination conditions were favourable. Their spectrophotometric measurements suggested that the ice crystals which comprise noctilucent clouds are not spherical, and show a range of sizes.

Some workers have suggested a link between meteor activity and the peak occurrence times for noctilucent clouds in the Northern Hemisphere, during late June to early July. At this time, Earth runs through a stream of debris from Encke's Comet, which produces the Beta Taurid meteor shower. This shower occurs during daylight, and may only be detected using radio-observing methods. A later, more favourable, encounter with the stream provides the weak night-time Taurid shower in November. In principle, the enhanced flux of meteoric debris into the high atmosphere should increase the availability of condensation nuclei for noctilucent cloud formation. Examination of noctilucent cloud reports collected over the years, however, shows that the main influence appears to be solar, and consequent geomagnetic, activity. When the Sun is active (as in the early 1990s), fewer displays are expected, since heating of the upper atmosphere during auroral activity should make conditions unfavourable for noctilucent cloud condensation. This does, indeed, seem to be the case, albeit with a slight lag (some have suggested two years) between peak sunspot and minimum noctilucent cloud occurrence.

The issue is confused somewhat by the observation, in the summer of 1987, of noctilucent clouds and aurora occurring *together* by Mark Zalcik and other Canadian observers. Duncan Waldron, a photographer at the Royal Observatory Edinburgh, recorded a similar simultaneous occurrence of the two phenomena from Scotland on July 13–14, 1991. Obviously, further observations are required to clarify the effects of auroral activity on noctilucent clouds: for example, does the onset of aurora disrupt an existing noctilucent cloud field?

Further recent work has brought the suggestion that noctilucent clouds may be sensitive indicators of long-term – possibly man-made – changes in the atmosphere as a whole. American workers have pointed out that early observations of noctilucent clouds, from about a century ago, were few and far between, and usually described only rather faint displays. Modern observers see more frequent, and brighter, noctilucent clouds! On a shorter timescale, work by Michael Gadsden at Aberdeen University indicates a rough doubling in the frequency with which observers in north-west Europe have reported noctilucent clouds in the past thirty years.

It has been proposed that this apparent increase in brightness and frequency is a result of increased liberation of methane into the atmosphere by industrial activities such as coal mining and oil exploration. Rising into the stratosphere, this methane is dissociated by the action of solar ultraviolet. A side-product of this reaction is hydrogen, which can then combine with hydroxyl radicals (OH) to produce water. G. E. Thomas and his co-workers in America suggest that two molecules of water should be produced for each methane molecule broken down. This model also provides a further source for the traces of water vapour which condense to form noctilucent clouds. Again, further observations are required to determine whether the apparent secular increase in noctilucent cloud frequency is genuine, and, if so, whether it will continue in years ahead. An interesting aside to all this is the further possibility that the noctilucent cloud 'season' may be getting longer, with Scottish sightings being made earlier into May and later into August than before.

Noctilucent clouds, then, are of more than just æsthetic interest. They are subject to a considerable amount of professional scientific study. In such studies, the professionals have found that the best collections of observational statistics on the appearance of noctilucent clouds have actually been accumulated by amateur astronomers! Such work was, initially, co-ordinated in the British Isles by the Balfour Stewart Laboratory of Edinburgh University, under the direction of James Paton. Since the Laboratory's closure in the 1970s, this work has continued under the umbrella of the British Astronomical Association Aurora Section. Indeed, the BAA group has extended the sphere of influence to encompass much of north-west Europe, and now collects data from Scandinavian observers. An independent network has also recently been established in Canada and the northern United States. Unfortunately, noctilucent clouds are seldom observed in the Southern Hemisphere, where the populated land-masses are poorly disposed with respect to the main zones of noctilucent cloud occurrence.

Observations are easy to make, and may be carried out with little more than the naked eye. The northern sky should be regularly checked on all possible clear nights during the summer. Figure 3 shows the 'windows' during which noctilucent clouds may be seen (in other words, the times of night when the Sun lies between 6° and 16° below the horizon) from latitudes 50°, 53°, and 56°N. As can be seen, the further north one goes, the fewer the hours of summer darkness – from Edinburgh, for example, the sky does not become properly dark at all between May 12 and August 6.

Noctilucent clouds give themselves away by appearing bright against the twilight, while tropospheric clouds are dark, and may even be seen in contrast against them. Beware of misidentifying cirrus overhead with noctilucent clouds; scattering of the twilight can sometimes make this appear to shine a diffuse, dull white. A good rule of thumb is that if there is bright cloud overhead, but not in the low northern sky near Capella, then cirrus is being observed! Identification of noctilucent clouds can easily be made using a pair of binoculars: the fine banding of noctilucent clouds bears magnification, while cirrus dissolves into a featureless haze. The silvery-blue colour of noctilucent clouds is also distinctive – and once seen, is unlikely to be forgotten by the observer. Towards the horizon, noctilucent clouds appear more golden in colour, as a result of atmospheric absorption.

Noctilucent cloud particles predominantly scatter sunlight forwards, and the cloudfield therefore appears brightest in the low, northern part of the sky: if present overhead and to the south,

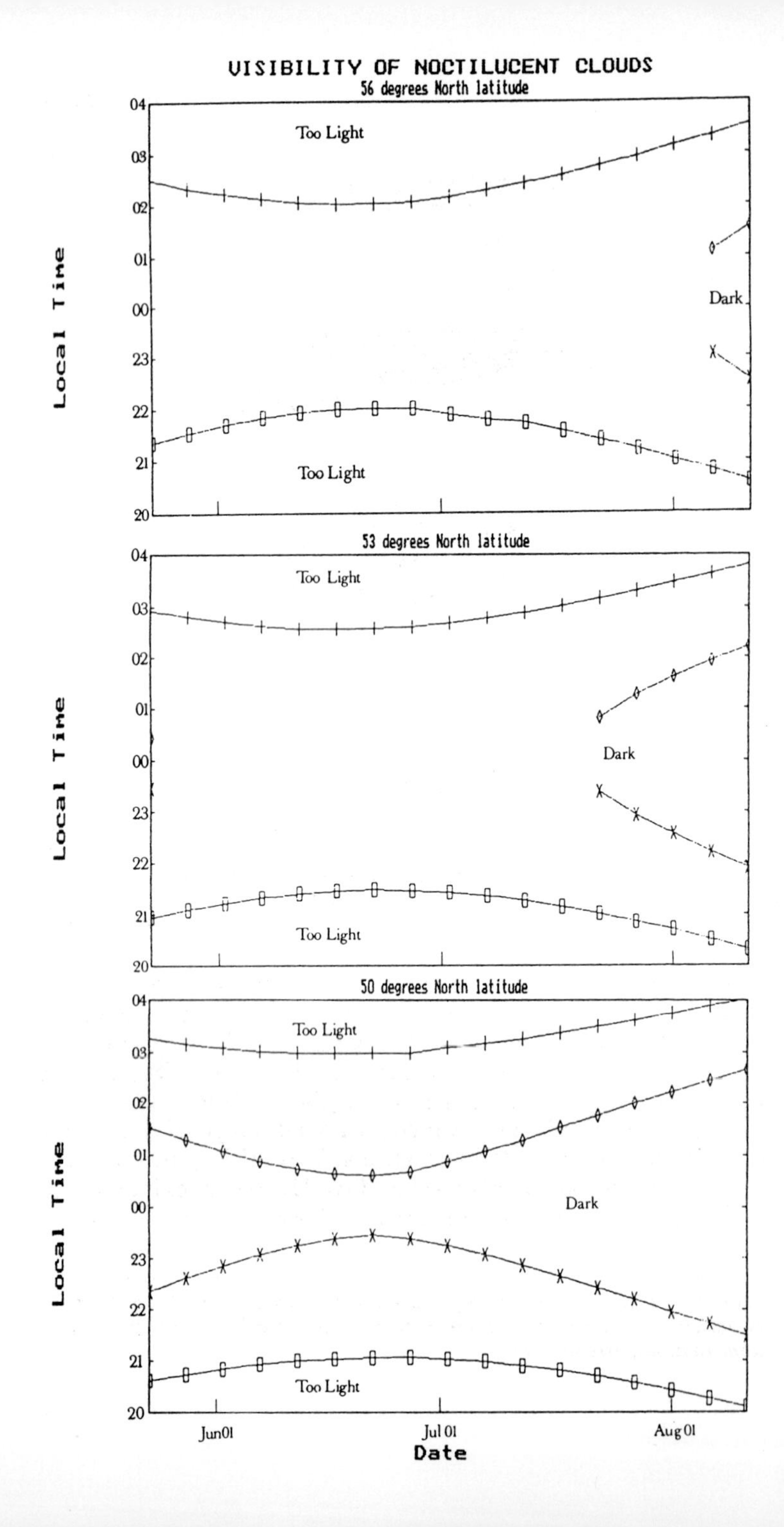

VISIBILITY OF NOCTILUCENT CLOUDS
56 degrees North latitude
Too Light
Dark
Too Light
53 degrees North latitude
Too Light
Dark
Too Light
50 degrees North latitude
Too Light
Dark
Too Light
Local Time
04
03
02
01
00
23
22
21
20
Jun01
Jul 01
Aug 01
Date

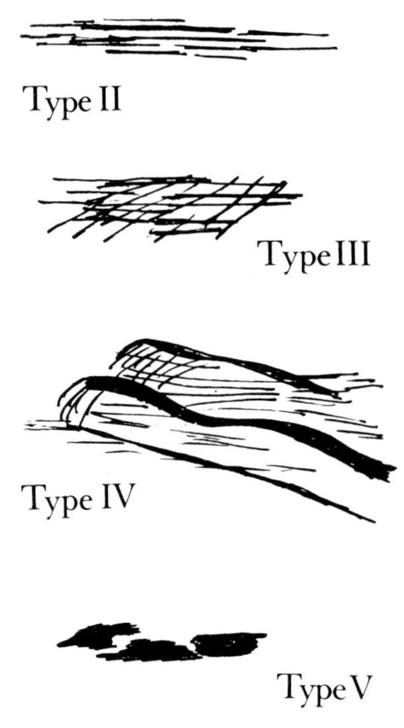

Figure 4. Schematic representation of the typical forms shown by noctilucent clouds. Type I, not shown here, is a featureless background veil.

noctilucent clouds appear faint in those parts of the sky. The brightest area in the cloudfield is usually immediately above the Sun's (hidden!) position below the horizon, and moves gradually around the horizon from north-west to north-east. Displays are most extensive just after sunset, and again just before dawn.

Noctilucent clouds show a relatively small range of *forms*, diagrammatically represented in Figure 4. Type II is quite common, consisting of roughly parallel bands. Type III comprises interwoven bands. Curved, feathery structures – *billows* – are often found in

---

Figure 3. Noctilucent clouds are only seen when the Sun lies between 6° and 16° below the horizon. Shown here are the 'windows' when these conditions prevail during the northern hemisphere summer.

brighter, more extensive displays, and are defined as Type IV. Sometimes, denser amorphous *patches* of Type V noctilucent cloud are seen. Displays may also be accompanied by a diffuse background veil (Type I, not illustrated) which will often be missed if it occurs in the absence of other forms.

The most basic useful observation is an indication that noctilucent clouds were visible at a given date and time from a certain location. More experienced observers can give a description of the types present, and the extent of these in altitude and azimuth. The brightness of various features may also be recorded, on an ascending scale of 1 to 3.

Such observations, made at 15-minute intervals, can be combined with others made elsewhere at the same time to assess the geographical extent of the cloudfield. Where possible, observations should be taken at standard times – on the hour, quarter-past, half-past, and quarter-to. Visual observations may usefully be complemented by annotated sketches of the clouds' appearance.

More accurate, still, from the point of view of obtaining measurements, are photographs. In addition to providing a scientifically-useful record, photographs of noctilucent clouds can be very appealing visually. The general light-level is usually sufficient for recording of foreground objects, enhancing the pictures.

Noctilucent clouds are best photographed in colour, and many observers prefer slide films. The author has enjoyed much success over the years using Kodak Ektachrome: exposures at ISO 400 on this emulsion are, typically, between 1–3 seconds at *f*/2.8, using a 'standard' 50-mm lens. Wide-angle lenses may sometimes be useful for recording extensive displays. The main factor which determines exposure time is the brightness of the sky background: brighter skies demand shorter exposures.

Both visual and photographic observations are, obviously, of greatest value if reported promptly to the appropriate co-ordinating body. In north-west Europe, results should be sent to the BAA Aurora Section: Dr D. Gavine, 29 Coillesdene Crescent, Edinburgh, EH15 2JJ, United Kingdom. North American reports may be submitted to: Mark Zalcik, #2 14225-82 Street, Edmonton, Alberta, T5 E 2V7, Canada.

Little known a century ago, and only partially understood even now, noctilucent clouds provide an occasional interesting, and sometimes rather beautiful, spectacle for observers during the high temperate latitude summer. With increasing scientific interest –

Figure 5. A fine display of noctilucent clouds was widely observed from central Scotland on June 30–July 1, 1983. This patch of Type IV noctilucent cloud was simultaneously photographed by Dave Gavine in Edinburgh and Alastair Simmons near Glasgow, allowing its precise geographical location and movement to be determined by trigonometry. This photograph was taken by Neil Bone, also from Edinburgh, at 23.41 UT.

both amateur and professional – in the phenomenon, there is less excuse than ever for observers to look upon the twilit summer evenings as an unwelcome, enforced lay-off period!

References

Bone, N. *The Aurora: Sun–Earth Interactions*. Ellis Horwood (1991).

Gadsden, M. and Schröder, W. *Noctilucent Clouds*. Springer-Verlag (1989).
Gavine, D. 'Noctilucent clouds over western Europe during 1989', *Meteorological Magazine* **120**, 65–66 (1991).
Soberman, R. K. 'Noctilucent clouds', *Scientific American* **208**, *5*, 84–96 (1963).
Thomas, G. E. *et al*. 'Relation between increasing methane and the presence of ice clouds at the mesopause', *Nature* **338**, 490–492 (1989).

Ripples in the Universe

PATRICK MOORE

In April 1992 – just as this *Yearbook* was going to press – came the announcement of a spectacular discovery. One of the world's greatest cosmologists, Stephen Hawking, even claimed that it was 'the discovery of the century, if not of all time'. This does seem to be an exaggeration, but there is no doubt that the revelation is extremely important.

According to modern theory, the universe – space, time, matter, everything – began in a 'Big Bang', some time between 15,000 million and 20,000 million years ago. At first the universe was intensely hot. Because space was created at the same moment, we cannot say 'where' the Big Bang happened – it happened everywhere – and neither can we ask what happened 'before the Big Bang', because there was no 'before'. There was a period of rapid inflation; the universe became larger, and also cooled down. By now the overall temperature has fallen to a mere 2.7 degrees above absolute zero, absolute zero being the lowest temperature possible, −273 degrees Centigrade.

In 1965 two American radio astronomers, Arno Penzias and Robert Wilson, were using a special radio telescope when they detected a background 'hiss' at microwave wavelength. They could not at first identify it, and believed that it might be due to pigeon droppings in their telescope; but before long they realized that it represented the last manifestation of the Big Bang. The radiation was coming in from all directions all the time.

Yet surprisingly, it was absolutely uniform. This meant that the universe would have had to be 'smooth' during the early inflationary period. Then how could the present universe, which is emphatically not smooth, have evolved at all?

Efforts to detect irregularities in the microwave background have been negative until now. Success has finally come thanks to the artificial satellite COBE (Cosmic Background Explorer) which was launched especially for the purpose. It was put into an orbit which carries it round the Earth at an altitude of around 560 miles; by careful design it was hoped to produce temperature maps of the sky accurate to 0.0003 of a degree, far better than anything achieved before. Since it began operations in 1989, COBE has made more than 300,000,000 measurements.

It has now revealed that there really are temperature variations in the background radiation. This in turn proves that there are differences in the density of matter at the very edge of the observable universe, and it seems that we have identified what may be called 'ripples'.

This is immensely significant. If the spreading out of matter had been as uniform as it seemed to be before the COBE results, there could have been no conceivable way in which galaxies could have formed; but with uneven distribution of density, gravitational forces could come into play, so that clumps of matter could draw together to produce galaxies. These in turn produced stars and planets, such as our own Earth.

What COBE cannot do, of course, is to explain the reason for the Big Bang itself. This remains as much of a mystery as ever. But at least we can now build up a coherent picture of what has happened since that one moment of creation.

Some Interesting Variable Stars

JOHN ISLES

The following stars are of interest for many reasons. Of course, the periods and ranges of many variables are not constant from one cycle to another.

Star	*R.A.*		*Declination*		*Range*	*Type*	*Period*	*Spectrum*
	h	*m*	*deg.*	*min.*			*days*	
R Andromedæ	00	24.0	+38	35	5.8–14.9	Mira	409	S
W Andromedæ	02	17.6	+44	18	6.7–14.6	Mira	396	S
U Antliæ	10	35.2	−39	34	5–6	Irregular	–	C
Theta Apodis	14	05.3	−76	48	5–7	Semi-regular	119	M
R Aquarii	23	43.8	−15	17	5.8–12.4	Symbiotic	387	M+Pec
T Aquarii	20	49.9	−05	09	7.2–14.2	Mira	202	M
R Aquilæ	19	06.4	+08	14	5.5–12.0	Mira	284	M
V Aquilæ	19	04.4	−05	41	6.6– 8.4	Semi-regular	353	C
Eta Aquilæ	19	52.5	+01	00	3.5– 4.4	Cepheid	7.2	F–G
U Aræ	17	53.6	−51	41	7.7–14.1	Mira	225	M
R Arietis	02	16.1	+25	03	7.4–13.7	Mira	187	M
U Arietis	03	11.0	+14	48	7.2–15.2	Mira	371	M
R Aurigæ	05	17.3	+53	35	6.7–13.9	Mira	458	M
Epsilon Aurigæ	05	02.0	+43	49	2.9– 3.8	Algol	9892	F+B
R Boötis	14	37.2	+26	44	6.2–13.1	Mira	223	M
W Boötis	14	43.4	+26	32	4.7– 5.4	Semi-regular?	450?	M
X Camelopardalis	04	45.7	+75	06	7.4–14.2	Mira	144	K–M
R Cancri	08	16.6	+11	44	6.1–11.8	Mira	362	M
X Cancri	08	55.4	+17	14	5.6– 7.5	Semi-regular	195?	C
R Canis Majoris	07	19.5	−16	24	5.7– 6.3	Algol	1.1	F
S Canis Minoris	07	32.7	+08	19	6.6–13.2	Mira	333	M
R Canum Ven.	13	49.0	+39	33	6.5–12.9	Mira	329	M
R Carinæ	09	32.2	−62	47	3.9–10.5	Mira	309	M
S Carinæ	10	09.4	−61	33	4.5– 9.9	Mira	149	K–M
l Carinæ	09	45.2	−62	30	3.3– 4.2	Cepheid	35.5	F–K
Eta Carinæ	10	45.1	−59	41	−0.8– 7.9	Irregular	–	Pec
R Cassiopeiæ	23	58.4	+51	24	4.7–13.5	Mira	430	M
S Cassiopeiæ	01	19.7	+72	37	7.9–16.1	Mira	612	S
W Cassiopeiæ	00	54.9	+58	34	7.8–12.5	Mira	406	C
Gamma Cass.	00	56.7	+60	43	1.6– 3.0	Irregular	–	B
Rho Cassiopeiæ	23	54.4	+57	30	4.1– 6.2	Semi-regular	–	F–K
R Centauri	14	16.6	−59	55	5.3–11.8	Mira	546	M
S Centauri	12	24.6	−49	26	7–8	Semi-regular	65	C
T Centauri	13	41.8	−33	36	5.5– 9.0	Semi-regular	90	K–M
S Cephei	21	35.2	+78	37	7.4–12.9	Mira	487	C
T Cephei	21	09.5	+68	29	5.2–11.3	Mira	388	M
Delta Cephei	22	29.2	+58	25	3.5– 4.4	Cepheid	5.4	F–G
Mu Cephei	21	43.5	+58	47	3.4– 5.1	Semi-regular	730	M
U Ceti	02	33.7	−13	09	6.8–13.4	Mira	235	M
W Ceti	00	02.1	−14	41	7.1–14.8	Mira	351	S
Omicron Ceti	02	19.3	−02	59	2.0–10.1	Mira	332	M

Star	*R.A.* *h*	*m*	*Declination* *deg.*	*min.*	*Range*	*Type*	*Period* *days*	*Spectrum*
R Chamæleontis	08	21.8	−76	21	7.5–14.2	Mira	335	M
T Columbæ	05	19.3	−33	42	6.6–12.7	Mira	226	M
R Comæ Ber.	12	04.3	+18	47	7.1–14.6	Mira	363	M
R Coronæ Bor.	15	48.6	+28	09	5.7–14.8	R Coronæ Bor.	–	C
S Coronæ Bor.	15	21.4	+31	22	5.8–14.1	Mira	360	M
T Coronæ Bor.	15	59.6	+25	55	2.0–10.8	Recurr. nova	–	M+Pec
V Coronæ Bor.	15	49.5	+39	34	6.9–12.6	Mira	358	C
W Coronæ Bor.	16	15.4	+37	48	7.8–14.3	Mira	238	M
R Corvi	12	19.6	−19	15	6.7–14.4	Mira	317	M
R Crucis	12	23.6	−61	38	6.4– 7.2	Cepheid	5.8	F–G
R Cygni	19	36.8	+50	12	6.1–14.4	Mira	426	S
U Cygni	20	19.6	+47	54	5.9–12.1	Mira	463	C
W Cygni	21	36.0	+45	22	5.0– 7.6	Semi-regular	131	M
RT Cygni	19	43.6	+48	47	6.0–13.1	Mira	190	M
SS Cygni	21	42.7	+43	35	7.7–12.4	Dwarf nova	50±	K+Pec
CH Cygni	19	24.5	+50	14	5.6– 9.0	Symbiotic	–	M+B
Chi Cygni	19	50.6	+32	55	3.3–14.2	Mira	408	S
R Delphini	20	14.9	+09	05	7.6–13.8	Mira	285	M
U Delphini	20	45.5	+18	05	5.6– 7.5	Semi-regular	110?	M
EU Delphini	20	37.9	+18	16	5.8– 6.9	Semi-regular	60	M
Beta Doradus	05	33.6	−62	29	3.5– 4.1	Cepheid	9.8	F–G
R Draconis	16	32.7	+66	45	6.7–13.2	Mira	246	M
T Eridani	03	55.2	−24	02	7.2–13.2	Mira	252	M
R Fornacis	02	29.3	−26	06	7.5–13.0	Mira	389	C
R Geminorum	07	07.4	+22	42	6.0–14.0	Mira	370	S
U Geminorum	07	55.1	+22	00	8.2–14.9	Dwarf nova	105±	Pec+M
Zeta Geminorum	07	04.1	+20	34	3.6– 4.2	Cepheid	10.2	F–G
Eta Geminorum	06	14.9	+22	30	3.2– 3.9	Semi-regular	233	M
S Gruis	22	26.1	−48	26	6.0–15.0	Mira	402	M
S Herculis	16	51.9	+14	56	6.4–13.8	Mira	307	M
U Herculis	16	25.8	+18	54	6.4–13.4	Mira	406	M
Alpha Herculis	17	14.6	+14	23	2.7– 4.0	Semi-regular	–	M
68, u Herculis	17	17.3	+33	06	4.7– 5.4	Algol	2.1	B+B
R Horologii	02	53.9	−49	53	4.7–14.3	Mira	408	M
U Horologii	03	52.8	−45	50	6–14	Mira	348	M
R Hydræ	13	29.7	−23	17	3.5–10.9	Mira	389	M
U Hydræ	10	37.6	−13	23	4.3– 6.5	Semi-regular	450?	C
VW Hydri	04	09.1	−71	18	8.4–14.4	Dwarf nova	27±	Pec
R Leonis	09	47.6	+11	26	4.4–11.3	Mira	310	M
R Leonis Minoris	09	45.6	+34	31	6.3–13.2	Mira	372	M
R Leporis	04	59.6	−14	48	5.5–11.7	Mira	427	C
Y Libræ	15	11.7	−06	01	7.6–14.7	Mira	276	M
RS Libræ	15	24.3	−22	55	7.0–13.0	Mira	218	M
Delta Libræ	15	01.0	−08	31	4.9– 5.9	Algol	2.3	A
R Lyncis	07	01.3	+55	20	7.2–14.3	Mira	379	S
R Lyræ	18	55.3	+43	57	3.9– 5.0	Semi-regular	46?	M
RR Lyræ	19	25.5	+42	47	7.1– 8.1	RR Lyræ	0.6	A–F
Beta Lyræ	18	50.1	+33	22	3.3– 4.4	Eclipsing	12.9	B
U Microscopii	20	29.2	−40	25	7.0–14.4	Mira	334	M
U Monocerotis	07	30.8	−09	47	5.9– 7.8	RV Tauri	91	F–K
V Monocerotis	06	22.7	−02	12	6.0–13.9	Mira	340	M
R Normæ	15	36.0	−49	30	6.5–13.9	Mira	508	M
T Normæ	15	44.1	−54	59	6.2–13.6	Mira	241	M
R Octantis	05	26.1	−86	23	6.3–13.2	Mira	405	M
S Octantis	18	08.7	−86	48	7.2–14.0	Mira	259	M
V Ophiuchi	16	26.7	−12	26	7.3–11.6	Mira	297	C
X Ophiuchi	18	38.3	+08	50	5.9– 9.2	Mira	329	M
RS Ophiuchi	17	50.2	−06	43	4.3–12.5	Recurr. nova	–	OB+M
U Orionis	05	55.8	+20	10	4.8–13.0	Mira	368	M
W Orionis	05	05.4	+01	11	5.9– 7.7	Semi-regular	212	C

Star	*R.A.*		*Declination*		*Range*	*Type*	*Period*	*Spectrum*
	h	*m*	*deg.*	*min.*			*days*	
Alpha Orionis	05	55.2	+07	24	0.0– 1.3	Semi-regular	2335	M
S Pavonis	19	55.2	−59	12	6.6–10.4	Semi-regular	381	M
Kappa Pavonis	18	56.9	−67	14	3.9– 4.8	Cepheid	9.1	G
R Pegasi	23	06.8	+10	33	6.9–13.8	Mira	378	M
Beta Pegasi	23	03.8	+28	05	2.3– 2.7	Irregular	—	M
X Persei	03	55.4	+31	03	6.0– 7.0	Gamma Cass.	–	09.5
Beta Persei	03	08.2	+40	57	2.1– 3.4	Algol	2.9	B
Rho Persei	03	05.2	+38	50	3.3– 4.0	Semi-regular	50?	M
Zeta Phœnicis	01	08.4	−55	15	3.9– 4.4	Algol	1.7	B+B
R Pictoris	04	46.2	−49	15	6.4–10.1	Semi-regular	171	M
L^2 Puppis	07	13.5	−44	39	2.6– 6.2	Semi-regular	141	M
T Pyxidis	09	04.7	−32	23	6.5–15.3	Recurr. nova	7000±	Pec
U Sagittæ	19	18.8	+19	37	6.5– 9.3	Algol	3.4	B+G
WZ Sagittæ	20	07.6	+17	42	7.0–15.5	Dwarf nova	11900±	A
R Sagittarii	19	16.7	−19	18	6.7–12.8	Mira	270	M
RR Sagittarii	19	55.9	−29	11	5.4–14.0	Mira	336	M
RT Sagittarii	20	17.7	−39	07	6.0–14.1	Mira	306	M
RU Sagittarii	19	58.7	−41	51	6.0–13.8	Mira	240	M
RY Sagittarii	19	16.5	−33	31	5.8–14.0	R Coronæ Bor.	–	G
RR Scorpii	16	56.6	−30	35	5.0–12.4	Mira	281	M
RS Scorpii	16	55.6	−45	06	6.2–13.0	Mira	320	M
RT Scorpii	17	03.5	−36	55	7.0–15.2	Mira	449	S
S Sculptoris	00	15.4	−32	03	5.5–13.6	Mira	363	M
R Scuti	18	47.5	−05	42	4.2– 8.6	RV Tauri	146	G–K
R Serpentis	15	50.7	+15	08	5.2–14.4	Mira	356	M
S Serpentis	15	21.7	+14	19	7.0–14.1	Mira	372	M
T Tauri	04	22.0	+19	32	9.3–13.5	Irregular	–	F–K
SU Tauri	05	49.1	+19	04	9.1–16.9	R Coronæ Bor.	–	G
Lambda Tauri	04	00.7	+12	29	3.4– 3.9	Algol	4.0	B+A
R Trianguli	02	37.0	+34	16	5.4–12.6	Mira	267	M
R Ursæ Majoris	10	44.6	+68	47	6.5–13.7	Mira	302	M
T Ursæ Majoris	12	36.4	+59	29	6.6–13.5	Mira	257	M
U Ursæ Minoris	14	17.3	+66	48	7.1–13.0	Mira	331	M
R Virginis	12	38.5	+06	59	6.1–12.1	Mira	146	M
S Virginis	13	33.0	−07	12	6.3–13.2	Mira	375	M
SS Virginis	12	25.3	+00	48	6.0– 9.6	Semi-regular	364	C
R Vulpeculæ	21	04.4	+23	49	7.0–14.3	Mira	137	M
Z Vulpeculæ	19	21.7	+25	34	7.3– 8.9	Algol	2.5	B+A

Mira Stars: maxima, 1993

JOHN ISLES

Below are given predicted dates of maxima for Mira stars that reach magnitude 7.5 or brighter at an average maximum. Individual maxima can in some cases be brighter or fainter than average by a magnitude or more, and all dates are only approximate. The positions, extreme ranges and mean periods of these stars can all be found in the preceding list of interesting variable stars.

Star	*Mean magnitude at max.*	*Dates of maxima*
R Andromedæ	6.9	Nov. 25
W Andromedæ	7.4	Mar. 1
R Aquarii	6.5	Jan. 14
R Aquilæ	6.1	Apr. 26
R Boötis	7.2	Aug. 4
R Cancri	6.8	Nov. 2
S Canis Minoris	7.5	Aug. 14
R Carinæ	4.6	Mar. 24
S Carinæ	5.7	Jan. 23, June 22, Nov. 18
R Cassiopeiæ	7.0	Aug. 28
R Centauri	5.8	Feb. 24, Oct. 6
T Cephei	6.0	Nov. 22
U Ceti	7.5	July 13
Omicron Ceti	3.4	June 19
T Columbæ	7.5	July 3
S Coronæ Borealis	7.3	Oct. 18
V Coronæ Borealis	7.5	May 21
R Corvi	7.5	May 10
R Cygni	7.5	Nov. 24
U Cygni	7.2	July 1
RT Cygni	7.3	June 4, Dec. 11
Chi Cygni	5.2	Apr. 25
R Geminorum	7.1	Oct. 23
U Herculis	7.5	Mar. 12
R Horologii	6.0	May 13
U Horologii	7	Mar. 23
R Hydræ	4.5	Jan. 7
R Leonis	5.8	July 12
R Leonis Minoris	7.1	Aug. 4

Star	*Mean magnitude at max.*	*Dates of maxima*
R Leporis	6.8	Feb. 8
RS Libræ	7.5	May 5, Dec. 9
V Monocerotis	7.0	Jan. 3, Dec. 9
R Normæ	7.2	Aug. 13
T Normæ	7.4	Jan. 9, Sep. 7
V Ophiuchi	7.5	Sep. 14
X Ophiuchi	6.8	Mar. 1
U Orionis	6.3	Nov. 15
R Sagittarii	7.3	Jan. 14, Oct. 11
RR Sagittarii	6.8	May 3
RT Sagittarii	7.0	May 22
RU Sagittarii	7.2	May 9
RR Scorpii	5.9	Mar. 15, Dec. 21
RS Scorpii	7.0	May 22
S Sculptoris	6.7	Nov. 4
R Serpentis	6.9	Mar. 30
R Trianguli	6.2	Sep. 6
R Ursæ Majoris	7.5	Aug. 4
R Virginis	6.9	Mar. 20, Aug. 13
S Virginis	7.0	Aug. 15

Some Interesting Double Stars

R. W. ARGYLE

Name	Magnitudes	Separation in seconds of arc	Position angle, degrees	Remarks
Gamma Andromedæ	2.3, 5.0	9.4	064	Yellow, blue. B is again double
Zeta Aquarii	4.3, 4.5	1.9	217	Slowly widening.
Gamma Arietis	4.8, 4.8	7.8	000	Very easy. Both white.
Theta Aurigæ	2.6, 7.1	3.5	313	Stiff test for 3-in.
Delta Boötis	3.5, 8.7	105	079	Fixed.
Epsilon Boötis	2.5, 4.9	2.8	335	Yellow, blue. Fine pair.
Zeta Cancri	5.6, 6.2	5.8	085	A again double.
Iota Cancri	4.2, 6.6	31	307	Easy. Yellow, blue.
Alpha Canum Ven.	2.9, 5.5	19.6	228	Easy. Yellow, bluish.
Alpha Capricorni	3.6, 4.2	379	291	Naked-eye pair.
Eta Cassiopeiæ	3.4, 7.5	12.5	312	Easy. Creamy, bluish.
Beta Cephei	3.2, 7.9	14	250	Easy with a 3-in.
Delta Cephei	var, 7.5	41	192	Very easy.
Alpha Centauri	0.0, 1.2	19.7	214	Very easy. Period 80 years.
Xi Cephei	4.4, 6.5	6.3	270	White, blue.
Gamma Ceti	3.5, 7.3	2.9	294	Not too easy.
Alpha Circini	3.2, 8.6	15.7	230	PA slowly decreasing.
Zeta Corona Bor.	5.1, 6.0	6.3	305	PA slowly increasing.
Delta Corvi	3.0, 9.2	24	214	Easy with a 3-in.
Alpha Crucis	1.4, 1.9	4.2	114	Third star in a low power field.
Gamma Crucis	1.6, 6.7	124	024	Third star in a low power field.
Beta Cygni	3.1, 5.1	34.5	055	Glorious. Yellow, blue.
61 Cygni	5.2, 6.0	30	148	Nearby binary. Period 722 years.
Gamma Delphini	4.5, 5.5	9.6	268	Easy. Yellowish, greenish.
Nu Draconis	4.9, 4.9	62	312	Naked eye pair.
Alpha Geminorum	1.9, 2.9	3.0	77	Widening. Visible with a 3-in.
Delta Geminorum	3.5, 8.2	6.5	120	Not too easy.
Alpha Herculis	var, 5.4	4.6	106	Red, green.
Delta Herculis	3.1, 8.2	9.8	272	Optical pair. Distance increasing.
Zeta Herculis	2.9, 5.5	1.6	084	Fine, rapid binary. Period 34 years.
Gamma Leonis	2.2, 3.5	4.4	123	Binary, 619 years.
Alpha Lyræ	0.0, 9.5	71	180	Optical pair. B is faint.

Name	*Magnitudes*	*Separation in seconds of arc*	*Position angle, degrees*	*Remarks*
{ Epsilon[1] Lyræ	5.0, 6.1	2.6	356	Quadruple system. Both
{ Epsilon[2] Lyræ	5.2, 5.5	2.2	093	pairs visible in a 3-in.
Zeta Lyræ	4.3, 5.9	44	149	Fixed. Easy double.
70 Ophiuchi	4.2, 6.0	1.5	224	Rapid motion.
Beta Orionis	0.1, 6.8	9.5	202	Can be split with a 3-in.
Iota Orionis	2.8, 6.9	11.8	141	Enmeshed in nebulosity.
Theta Orionis	6.7, 7.9	8.7	032	Trapezium in M42.
	5.1, 6.7	13.4	061	
Sigma Orionis	4.0, 10.3	11.4	238	Quintuple. A is a
	6.5, 7.5	30.1	231	close double.
Zeta Orionis	1.9, 4.0	2.4	162	Can be split in 3-in.
Eta Persei	3.8, 8.5	28.5	300	Yellow, bluish.
Beta Phœnicis	4.0, 4.2	1.5	324	Slowly widening.
Beta Piscis Aust.	4.4, 7.9	30.4	172	Optical pair. Fixed.
Alpha Piscium	4.2, 5.1	1.9	278	Binary, 933 years.
Kappa Puppis	4.5, 4.7	9.8	318	Both white.
Alpha Scorpii	1.2, 5.4	3.0	275	Red, green.
Nu Scorpii	4.3, 6.4	42	336	Both again double.
Theta Serpentis	4.5, 5.4	22.3	103	Fixed. Very easy.
Alpha Tauri	0.9, 11.1	131	032	Wide, but B very faint in small telescopes.
Beta Tucanæ	4.4, 4.8	27.1	170	Both again double.
Zeta Ursæ Majoris	2.3, 4.0	14.4	151	Very easy. Naked eye pair with Alcor.
Xi Ursæ Majoris	4.3, 4.8	1.3	060	Binary, 60 years. Closing. Needs a 4-in.
Gamma Virginis	3.5, 3.5	3.0	287	Binary, 171 years. Closing.
Theta Virginis	4.4, 9.4	7.1	343	Not too easy.
Gamma Volantis	3.9, 5.8	13.8	299	Very slow binary.

Some Interesting Nebulæ and Clusters

Object	R.A.		Dec.		*Remarks*
	h	*m*			
M.31 Andromedæ	00	40.7	+41	05	Great Galaxy, visible to naked eye.
H.VIII 78 Cassiopeiæ	00	41.3	+61	36	Fine cluster, between Gamma and Kappa Cassiopeiæ.
M.33 Trianguli	01	31.8	+30	28	Spiral. Difficult with small apertures.
H.VI 33–4 Persei	02	18.3	+56	59	Double cluster; Sword-handle.
△142 Doradûs	05	39.1	−69	09	Looped nebula round 30 Doradûs. Naked-eye. In Large Cloud of Magellan.
M.1 Tauri	05	32.3	+22	00	Crab Nebula, near Zeta Tauri.
M.42 Orionis	05	33.4	−05	24	Great Nebula. Contains the famous Trapezium, Theta Orionis.
M.35 Geminorum	06	06.5	+24	21	Open cluster near Eta Geminorum.
H.VII 2 Monocerotis	06	30.7	+04	53	Open cluster, just visible to naked eye.
M.41 Canis Majoris	06	45.5	−20	42	Open cluster, just visible to naked eye.
M.47 Puppis	07	34.3	−14	22	Mag. 5,2. Loose cluster.
H.IV 64 Puppis	07	39.6	−18	05	Bright planetary in rich neighbourhood.
M.46 Puppis	07	39.5	−14	42	Open cluster.
M.44 Cancri	08	38	+20	07	Præsepe. Open cluster near Delta Cancri. Visible to naked eye.
M.97 Ursæ Majoris	11	12.6	+55	13	Owl Nebula, diameter 3′. Planetary.
Kappa Crucis	12	50.7	−60	05	'Jewel Box'; open cluster, with stars of contrasting colours.
M.3 Can. Ven.	13	40.6	+28	34	Bright globular.
Omega Centauri	13	23.7	−47	03	Finest of all globulars. Easy with naked eye.
M.80 Scorpii	16	14.9	−22	53	Globular, between Antares and Beta Scorpionis.
M.4 Scorpii	16	21.5	−26	26	Open cluster close to Antares.
M.13 Herculis	16	40	+36	31	Globular. Just visible to naked eye.
M.92 Herculis	16	16.1	+43	11	Globular. Between Iota and Eta Herculis.
M.6 Scorpii	17	36.8	−32	11	Open cluster; naked eye.
M.7 Scorpii	17	50.6	−34	48	Very bright open cluster; naked eye.
M.23 Sagittarii	17	54.8	−19	01	Open cluster nearly 50′ in diameter.
H.IV 37 Draconis	17	58.6	+66	38	Bright planetary.
M.8 Sagittarii	18	01.4	−24	23	Lagoon Nebula. Gaseous. Just visible with naked eye.
NGC 6572 Ophiuchi	18	10.9	+06	50	Bright planetary, between Beta Ophiuchi and Zeta Aquilæ.
M.17 Sagittarii	18	18.8	−16	12	Omega Nebula. Gaseous. Large and bright.
M.11 Scuti	18	49.0	−06	19	Wild Duck. Bright open cluster.
M.57 Lyræ	18	52.6	+32	59	Ring Nebula. Brightest of planetaries.
M.27 Vulpeculæ	19	58.1	+22	37	Dumb-bell Nebula, near Gamma Sagittæ.

Object	R.A.		Dec.		*Remarks*
	h	*m*			
H.IV 1 Aquarii	21	02.1	−11	31	Bright planetary near Nu Aquarii.
M.15 Pegasi	21	28.3	+12	01	Bright globular, near Epsilon Pegasi.
M.39 Cygni	21	31.0	+48	17	Open cluster between Deneb and Alpha Lacertæ. Well seen with low powers.

Our Contributors

Dr David Allen needs no introduction to *Yearbook* readers; he continues his work at the Anglo-Australian Observatory in New South Wales, and is at present Acting Director.

Neil Bone is the Director of the Meteor Section of the British Astronomical Association, but also has a special interest in auroræ and other 'sky glows'.

John Isles lives in Cyprus, and is the Director of the Variable Star Section of the British Astronomical Association.

Dr Mark Kidger is also an astrophysicist with a special interest in quasars, but is also an enthusiastic planetary observer. He carries out these researches at the observatory in Tenerife.

Dr Derek McNally is the Director of the University of London Observatory – and as such is particularly concerned with light pollution problems!

Colin Ronan is a well known astronomical historian; he has written many books, and has been President of the British Astronomical Association as well as serving for many years as the Editor of its journal.

Dr Jasper Wall, of the Royal Greenwich Observatory, is an astrophysicist who has made many important contributions, and was for some years in charge of all our telescopes at La Palma.

The William Herschel Society maintains the museum now established at 19 New King Street, Bath – the only surviving Herschel House. It also undertakes activities of various kinds. New members would be welcome; those interested are asked to contact Dr L. Hilliard at 2 Lambridge, London Road, Bath.

Astronomical Societies in Great Britain

British Astronomical Association
Assistant Secretary: Burlington House, Piccadilly, London W1V 9AG.
Meetings: Lecture Hall of Scientific Societies, Civil Service Commission Building, 23 Savile Row, London W1. Last Wednesday each month (Oct.–June). 1700 hrs and some Saturday afternoons.
Association for Astronomy Education
Secretary: Bob Kibble, 34 Ackland Crescent, Denmark Hill, London SE5 8EQ.
Astronomical Society of Wales
Secretary: John Minopoli, 12 Gwendoline Street, Port Talbot, West Glamorgan.
Federation of Astronomical Societies
Secretary: Mrs Christine Sheldon, Whitehaven, Lower Moor, Pershore, Worcs.
Junior Astronomical Society
Secretary: M. Ratcliffe, 36 Fairway, Keyworth, Nottingham.
Meetings: Central Library, Theobalds Road, London WC1. Last Saturday Jan., April, July, Oct. 2.30 p.m.
Junior Astronomical Society of Ireland
Secretary: K. Nolan, 5 St Patrick's Crescent, Rathcoole, Co. Dublin.
Meetings: The Royal Dublin Society, Ballsbridge, Dublin 4. Monthly.
Aberdeen and District Astronomical Society
Secretary: Stephen Graham, 25 Davidson Place, Northfield, Aberdeen.
Meetings: Robert Gordon's Institute of Technology, St Andrew's Street, Aberdeen. Friday 7.30 p.m.
Altrincham and District Astronomical Society
Secretary: Colin Henshaw, 10 Delamore Road, Gatley, Cheadle, Cheshire.
Meetings: Public Library, Timperley. 1st Friday of each month, 7.30 p.m.
Astra Astronomy Section
Secretary: Ian Downie, 151 Sword Street, Glasgow G31.
Meetings: Public Library, Airdrie. Weekly.
Aylesbury Astronomical Society
Secretary: Ian Welland, 20 Woodcote Green, Dounley, High Wycombe HP13 5UN.
Bassetlaw Astronomical Society
Secretary: P. R. Stanley, 28 Festival Avenue, Harworth, nr. Doncaster.
Meetings: Farr Community Hall, Chapel Walk, Westgate, Worksop, Notts. Tuesday fortnightly, 7.30 p.m.
Batley & Spenborough Astronomical Society
Secretary: Robert Morton, 22 Links Avenue, Cleckheaton, West Yorks BD19 4EG.
Meetings: Milner K. Ford Observatory, Wilton Park, Batley. Every Thursday, 7.30 p.m.
Bedford Astronomical Society
Secretary: D. Eagle, 24 Copthorne Close, Oakley, Bedford.
Meetings: Bedford School, Burnaby Rd, Bedford. Last Tuesday each month.
Bingham & Brookes Space Organization
Secretary: N. Bingham, 15 Hickmore's Lane, Lindfield, W. Sussex.
Birmingham Astronomical Society
Secretary: P. Truelove, 58 Taylor Road, King's Heath, Birmingham.
Meetings: Room 261, University of Aston, last Tuesday each month, Sept. to May.
Blackpool & District Astronomical Society
Secretary: J. L. Crossley, 24 Fernleigh Close, Bispham, Blackpool, Lancs.
Bolton Astronomical Society
Secretary: Peter Miskiw, 9 Hedley Street, Bolton.
Border Astronomical Society
Secretary: David Pettit, 14 Shap Grove, Carlisle, Cumbria.
Boston Astronomers
Secretary: B. Tongue, South View, Fen Road, Stickford, Boston.
Meetings: Details from the Secretary.
Bradford Astronomical Society
Secretary: John Schofield, Briar Lea, Bromley Road, Bingley, W. Yorks.
Meetings: Eccleshill Library, Bradford 2. Monday fortnightly (with occasional variations).
Braintree, Halstead & District Astronomical Society
Secretary: Heather Reeder, The Knoll, St Peters in the Field, Braintree, Essex.
Meetings: St Peter's Church Hall, St Peter's Road, Braintree, Essex. 3rd Thursday each month, 8 p.m.
Bridgend Amateur Astronomical Society
Secretary: J. M. Pugsley, 32 Hoel Fawr, Broadlands, North Cornelly, Bridgend.
Meetings: G.P. Room, Recreation Centre, Bridgend, 1st and 3rd Friday monthly, 7.30 p.m.

Bridgwater Astronomical Society
Secretary: W. L. Buckland, 104 Polden Street, Bridgwater, Somerset.
Meetings: Room D10, Bridgwater College, Bath Road Centre, Bridgwater. 2nd Wednesday each month, Sept.–June.

Brighton Astronomical Society
Secretary: Mrs B. C. Smith, Flat 2, 23 Albany Villas, Hove, Sussex BN3 2RS.
Meetings: Preston Tennis Club, Preston Drive, Brighton. Weekly, Tuesdays.

Bristol Astronomical Society
Secretary: Y. A. Sage, 33 Mackie Avenue, Filton, Bristol.
Meetings: Royal Fort (Rm G44), Bristol University. Every Friday each month, Sept.–May. Fortnightly, June–August.

Cambridge Astronomical Association
Secretary: R. J. Greening, 20 Cotts Croft, Great Chishill, Royston, Herts.
Meetings: Venues as published in newsletter. 1st and 3rd Friday each month, 8 p.m.

Cardiff Astronomical Society
Secretary: D. W. S. Powell, 1 Tal-y-Bont Road, Ely, Cardiff.
Meeting Place: Room 230, Dept. Law, University College, Museum Avenue, Cardiff. Alternate Thursdays, 8 p.m.

Castle Point Astronomy Club
Secretary: Miss Zena White, 43 Lambeth Road, Eastwood, Essex.
Meetings: St Michael's Church, Thundersley. Most Wednesdays, 8 p.m.

Chelmsford Astronomers
Secretary: Brendan Clark, 5 Borda Close, Chelmsford, Essex.
Meetings: Once a month.

Chelmsford and District Astronomical Society
Secretary: Miss C. C. Puddick, 6 Walpole Walk, Rayleigh, Essex.
Meetings: Sandon House School, Sandon, near Chelmsford. 2nd and last Monday of month. 7.45 p.m.

Chester Astronomical Society
Secretary: Mrs S. Brooks, 39 Halton Road, Great Sutton, South Wirral.
Meetings: Southview Community Centre, Southview Road, Chester. Last Monday each month except Aug. and Dec., 7.30 p.m.

Chester Society of Natural Science Literature and Art
Secretary: Paul Braid, 'White Wing', 38 Bryn Avenue, Old Colwyn, Colwyn Bay, Clwyd.
Meetings: Grosvenor Museum, Chester. Fortnightly.

Chesterfield Astronomical Society
Secretary: P. Lisewski, 148 Old Hall Road, Brampton, Chesterfield.
Meetings: Barnet Observatory, Newbold. Each Friday.

Clacton & District Astronomical Society
Secretary: C. L. Haskell, 105 London Road, Clacton-on-Sea, Essex.

Cleethorpes & District Astronomical Society
Secretary: C. Illingworth, 38 Shaw Drive, Grimsby, S. Humberside.
Meetings: Beacon Hill Observatory, Cleethorpes. 1st Wednesday each month.

Cleveland & Darlington Astronomical Society
Secretary: Neil Haggath, 5 Fountains Crescent, Eston, Middlesbrough, Cleveland.
Meetings: Elmwood Community Centre, Greens Lane, Hartburn, Stockton-on-Tees. Monthly, usually second Friday.

Colchester Amateur Astronomers
Secretary: F. Kelly, 'Middleton', Church Road, Elmstead Market, Colchester, Essex.
Meetings: William Loveless Hall, High Street, Wivenhoe. Friday evenings. Fortnightly.

Cotswold Astronomical Society
Secretary: Trevor Talbot, Innisfree, Winchcombe Road, Sedgebarrow, Worcs.
Meetings: Fortnightly in Cheltenham or Gloucester.

Coventry & Warwicks Astronomical Society
Secretary: Steve Payne, 73 Torbay Road, Allesley Park, Coventry.
Meetings: Coventry Technical College. 1st Friday each month, Sept.–June.

Crawley Astronomical Society
Secretary: G. Cowley, 67 Climpixy Road, Ifield, Crawley, Sussex.
Meetings: Crawley College of Further Education. Monthly Oct.–June.

Crayford Manor House Astronomical Society
Secretary: R. H. Chambers, Manor House Centre, Crayford, Kent.
Meetings: Manor House Centre, Crayford. Monthly during term-time.

Croydon Astronomical Society
Secretary: N. Fisher, 5 Dagmar Road, London SE25 6HZ.
Meetings: Lanfranc High School, Mitcham Road, Croydon. Alternate Fridays, 7.45 p.m.

Derby & District Astronomical Society
Secretary: Jane D. Kirk, 7 Cromwell Avenue, Findern, Derby.
Meetings: At home of Secretary. First and third Friday each month, 7.30 p.m.

Doncaster Astronomical Society
Secretary: J. A. Day, 297 Lonsdale Avenue, Intake, Doncaster.
Meetings: Fridays, weekly.
Dundee Astronomical Society
Secretary: G. Young, 37 Polepark Road, Dundee, Angus.
Meetings: Mills Observatory, Balgay Park, Dundee. First Friday each month, 7.30 p.m. Sept.–April.
Easington and District Astronomical Society
Secretary: T. Bradley, 52 Jameson Road, Hartlepool, Co. Durham.
Meetings: Easington Comprehensive School, Easington Colliery. Every third Thursday throughout the year, 7.30 p.m.
Eastbourne Astronomical Society
Secretary: P. Garter, 16 Redoubt Road, Eastbourne, East Sussex.
Meetings: St Aiden's Church Hall, 1 Whitley Road, Eastbourne. Monthly (except July and August).
East Lancashire Astronomical Society
Secretary: D. Chadwick, 16 Worston Lane, Great Harwood, Blackburn BB6 7TH.
Meetings: As arranged. Monthly.
Astronomical Society of Edinburgh
Secretary: R. G. Fenoulhet, 7 Greenend Gardens, Edinburgh EH17 7QB.
Meetings: City Observatory, Calton Hill, Edinburgh. Monthly.
Edinburgh University Astronomical Society
Secretary: c/o Dept. of Astronomy, Royal Observatory, Blackford Hill, Edinburgh.
Ewell Astronomical Society
Secretary: Edward Hanna, 91 Tennyson Avenue, Motspur Park, Surrey.
Meetings: 1st Friday of each month.
Exeter Astronomical Society
Secretary: Miss J. Corey, 5 Egham Avenue, Topsham Road, Exeter.
Meetings: The Meeting Room Wynards, Magdalen Street, Exeter. 1st Thursday of month.
Farnham Astronomical Society
Secretary: Laurence Anslow, 14 Wellington Lane, Farnham, Surrey.
Meetings: Church House, Union Road, Farnham. 2nd Monday each month, 7.45 p.m.
Fitzharry's Astronomical Society (Oxford & District)
Secretary: Mark Harman, 20 Lapwing Lane, Cholsey, Oxon.
Meetings: All Saints Methodist Church, Dorchester Crescent, Abingdon, Oxon.
Furness Astronomical Society
Secretary: A. Thompson, 52 Ocean Road, Walney Island, Barrow-in-Furness, Cumbria.
Meetings: St Mary's Church Centre, Dalton-in-Furness. 2nd Saturday in month, 7.30 p.m. No August meeting.
Fylde Astronomical Society
Secretary: 28 Belvedere Road, Thornton, Lancs.
Meetings: Stanley Hall, Rossendale Avenue South. 1st Wednesday each month.
Astronomical Society of Glasgow
Secretary: Malcolm Kennedy, 32 Cedar Road, Cumbernauld, Glasgow.
Meetings: University of Strathclyde, George St., Glasgow. 3rd Thursday each month, Sept.–April.
Greenock Astronomical Association
Secretary: Miss Fiona McKechnie, 19 Grey Place, Greenock.
Meetings: Greenock Arts Guild, 3 Campbell Street, Greenock.
Grimsby Astronomical Society
Secretary: R. Williams, 14 Richmond Close, Grimsby, South Humberside.
Meetings: Secretary's home. 2nd Thursday each month, 7.30 p.m.
Guernsey: La Société Guernesiaise Astronomy Section
Secretary: David Le Conte, Belle Etoile, Rue de Hamel, Castel, Guernsey.
Meetings: Monthly.
Guildford Association Society
Secretary: Mrs Joan Prosser, 115 Farnham Road, Guildford, Surrey.
Meetings: Guildford Institute, Ward Street, Guildford. 1st Thursday each month. Sept.–June, 7.30 p.m.
Gwynedd Astronomical Society
Secretary: P. J. Curtis, Ael-y-bryn, Malltraeth St Newborough, Anglesey, Gwynedd.
Meetings: Physics Lecture Room, Bangor University. 1st Thursday each month, 7.30 p.m.
The Hampshire Astronomical Group
Secretary: R. Dodd, 1 Conifer Close, Cowplain, Portsmouth.
Meetings: Clanfield Observatory. Each Friday, 7.30 p.m.
Astronomical Society of Haringey
Secretary: Wally Baker, 58 Stirling Road, Wood Green, London N22.
Meetings: The Hall of the Good Shepherd, Berwick Road, Wood Green. 3rd Wednesday each month, 8 p.m.
Harrogate Astronomical Society
Secretary: P. Barton, 31 Gordon Avenue, Harrogate, North Yorkshire.
Meetings: Harlow Hill Methodist Church Hall, 121 Otley Road, Harrogate. Last Friday each month.

Heart of England Astronomical Society
Secretary: R. D. Januszewski, 24 Emsworth Grove, Kings Heath, Birmingham.
Meetings: Chelmsley Wood Library. Last Thursday each month.

Hebden Bridge Literary & Scientific Society, Astronomical Section
Secretary: F. Parker, 48 Caldene Avenue, Mytholmroyd, Hebden Bridge, West Yorkshire.

Herschel Astronomy Society
Secretary: D. R. Whittaker, 149 Farnham Lane, Slough.
Meetings: Eton College, 2nd Friday each month.

Howards Astronomy Club
Secretary: H. Ilett, 22 St Georges Avenue, Warblington, Havant, Hants.
Meetings: To be notified.

Huddersfield Astronomical and Philosophical Society
Secretary (Assistant): M. Armitage, 37 Frederick Street, Crossland Moor, Huddersfield.
Meetings: 4A Railway Street, Huddersfield. Every Friday, 7.30 p.m.

Hull and East Riding Astronomical Society
Secretary: J. I. Booth, 3 Lynngarth Avenue, Cottingham, North Humberside.
Meetings: Ferens Recreation Centre, Chanterlands Avenue, Hull. 1st Friday each month, Oct.–April, 7.30 p.m.

Ilkeston & District Astronomical Society
Secretary: Trevor Smith, 129 Heanor Road, Smalley, Derbyshire.
Meetings: The Friends Meeting Room, Ilkeston Museum, Ilkeston. 2nd Tuesday monthly, 7.30 p.m.

Ipswich, Orwell Astronomical Society
Secretary: R. Gooding, 168 Ashcroft Road, Ipswich.
Meetings: Orwell Park Observatory, Nacton, Ipswich. Wednesdays 8 p.m.

Irish Astronomical Association
Secretary: Michael Duffy, 26 Ballymurphy Road, Belfast, Northern Ireland.
Meetings: Room 315, Ashby Institute, Stranmills Road, Belfast. Fortnightly. Wednesdays, Sept.–April, 7.30 p.m.

Irish Astronomical Society
Secretary: c/o PO Box 2547, Dublin 15, Eire.

Isle of Man Astronomical Society
Secretary: James Martin, Ballatepson Farm, Peel, Isle of Man.
Meetings: Falcon Cliff Hotel, Douglas, 1st Thursday each month, 8.30 p.m.

Isle of Wight Astronomical Society
Secretary: J. W. Feakins, 1 Hilltop Cottages, High Street, Freshwater, Isle of Wight.
Meetings: Unitarian Church Hall, Newport, Isle of Wight. Monthly.

Keele Astronomical Society
Secretary: Miss Caterina Callus, University of Keele, Keele, Staffs.
Meetings: As arranged during term time.

Kettering and District Astronomical Society
Asst. Secretary: Steve Williams, 120 Brickhill Road, Wellingborough, Northants.
Meetings: Quaker Meeting Hall, Northall Street, Kettering, Northants. 1st Tuesday each month. 7.45 p.m.

King's Lynn Amateur Astronomical Association
Secretary: P. Twynman, 17 Poplar Avenue, RAF Marham, King's Lynn.
Meetings: As arranged.

Lancaster and Morecambe Astronomical Society
Secretary: Miss E. Haygarth, 27 Coulston Road, Bowerham, Lancaster.
Meetings: Midland Hotel, Morecambe. 1st Wednesday each month except January. 7.30 p.m.

Lancaster University Astronomical Society
Secretary: c/o Students Union, Alexandra Square, University of Lancaster.
Meetings: As arranged.

Laymans Astronomical Society
Secretary: John Evans, 10 Arkwright Walk, The Meadows, Nottingham.
Meetings: The Popular, Bath Street, Ilkeston, Derbyshire. Monthly.

Leeds Astronomical Society
Secretary: A. J. Higgins, 23 Montagu Place, Leeds LS8 2RQ.
Meetings: Lecture Room, City Museum Library, The Headrow, Leeds.

Leicester Astronomical Society
Secretary: Dereck Brown, 64 Grange Drive, Glen Parva, Leicester.
Meetings: Judgemeadow Community College, Marydene Drive, Evington, Leicester. 2nd and 4th Tuesdays each month, 7.30 p.m.

Letchworth and District Astronomical Society
Secretary: Eric Hutton, 14 Folly Close, Hitchin, Herts.
Meetings: As arranged.

Limerick Astronomy Club
Secretary: Tony O'Hanlon, 26 Ballycannon Heights, Meelick, Co. Clare, Ireland.
Meetings: Limerick Senior College, Limerick, Ireland. Monthly (except June and August), 8 p.m.

Lincoln Astronomical Society
Secretary: G. Winstanley, 36 Cambridge Drive, Washingborough, Lincoln.
Meetings: The Lecture Hall, off Westcliffe Street, Lincoln. 1st Tuesday each month.

Liverpool Astronomical Society
Secretary: David Whittle, 17 Sandy Lane, Tuebrook, Liverpool.
Meetings: City Museum, Liverpool. Wednesdays and Fridays, monthly.

Loughton Astronomical Society
Meetings: Loughton Hall, Rectory Lane, Loughton, Essex. Thursdays 8 p.m.

Lowestoft and Great Yarmouth Regional Astronomers (LYRA) Society
Secretary: S. Briggs, 65 Stubbs Wood, Gunton Park, Lowestoft, Suffolk.
Meetings: Committee Room No. 30, Lowestoft College of F.E., St Peter's Street, Lowestoft. 3rd Thursday, Sept.–May (weather permitting on Corton Cliff site), 7.15 p.m.

Luton & District Astronomical Society
Secretary: D. Childs, 6 Greenways, Stopsley, Luton.
Meetings: Luton College of Higher Education, Park Square, Luton. Second and last Friday each month, 7.30 p.m.

Lytham St Annes Astronomical Association
Secretary: K. J. Porter, 141 Blackpool Road, Ansdell, Lytham St Annes, Lancs.
Meetings: College of Further Education, Clifton Drive South, Lytham St Annes. 2nd Wednesday monthly Oct.–June.

Macclesfield Astronomical Society
Secretary: Mrs C. Moss, 27 Westminster Road, Macclesfield, Cheshire.
Meetings: The Planetarium, Jodrell Bank, 1st Tuesday each month.

Maidenhead Astronomical Society
Secretary: c/o Chairman, Peter Hunt, Hightrees, Holyport Road, Bray, Berks.
Meetings: Library. Monthly (except July) 1st Friday.

Maidstone Astronomical Society
Secretary: N. O. Harris, 19 Greenside, High Hadden, Ashford, Kent.
Meetings: Nettlestead Village Hall, 1st Tuesday in month except July and Aug. 7.30 p.m.

Manchester Astronomical Society
Secretary: J. H. Davidson, Godlee Observatory, UMIST, Sackville Street, Manchester 1.
Meetings: At the Observatory, Thursdays, 7.30–9 p.m.

Mansfield and Sutton Astronomical Society
Secretary: G. W. Shepherd, Sherwood Observatory, Coxmoor Road, Sutton-in-Ashfield, Notts.
Meetings: Sherwood Observatory, Coxmoor Road. Last Tuesday each month, 7.55 p.m.

Mexborough and Swinton Astronomical Society
Secretary: Mark R. Benton, 61 The Lea, Swinton, Mexborough, Yorks.
Meetings: Methodist Hall, Piccadilly Road, Swinton, Near Mexborough. Thursdays, 7 p.m.

Mid-Kent Astronomical Society
Secretary: Brian A. van de Peep, 11 Berber Road, Strood, Rochester, Kent.
Meetings: Medway Teachers Centre, Vicarage Road, Strood, Rochester, Kent. Last Friday in month. Mid Kent College, Horsted. 2nd Friday in month.

Milton Keynes Astronomical Society
Secretary: The Secretary, Milton Keynes Astronomical Society, Bradwell Abbey Field Centre, Bradwell, Milton Keynes MK1 39AP.
Meetings: Alternate Tuesdays.

Moray Astronomical Society
Secretary: Richard Pearce, 1 Forsyth Street, Hopeman, Elgin, Moray, Scotland.
Meetings: Village Hall Close, Co. Elgin.

Newbury Amateur Astronomical Society
Secretary: Mrs A. Davies, 11 Sedgfield Road, Greenham, Newbury, Berks.
Meetings: United Reform Church Hall, Cromwell Road, Newbury. Last Friday of month, Aug.–May.

Newcastle-on-Tyne Astronomical Society
Secretary: C. E. Willits, 24 Acomb Avenue, Seaton Delaval, Tyne and Wear.
Meetings: Zoology Lecture Theatre, Newcastle University. Monthly.

North Aston Space & Astronomical Club
Secretary: W. R. Chadburn, 14 Oakdale Road, North Aston, Sheffield.
Meetings: To be notified.

Northamptonshire Natural History Astronomical Society
Secretary: Dr Nick Hewitt, 4 Daimler Close, Northampton.
Meetings: Humphrey Rooms, Castillian Terrace, Northampton. 2nd and last Monday each month.

North Devon Astronomical Society
Secretary: P. G. Vickery, 12 Broad Park Crescent, Ilfracombe, North Devon.
Meetings: Pilton Community College, Chaddiford Lane, Barnstaple. 1st Wednesday each month, Sept.–May.

North Dorset Astronomical Society
Secretary: J. E. M. Coward, The Pharmacy, Stalbridge, Dorset.
Meetings: Charterhay, Stourton, Caundle, Dorset. 2nd Wednesday each month.

North Staffordshire Astronomical Society
Secretary: N. Oldham, 25 Linley Grove, Alsager, Stoke-on-Trent.
Meetings: 1st Wednesday of each month at Cartwright House, Broad Street, Hanley.

North Western Association of Variable Star Observers
Secretary: Jeremy Bullivant, 2 Beaminster Road, Heaton Mersey, Stockport, Cheshire.
Meetings: Four annually.

Norwich Astronomical Society
Secretary: Malcolm Jones, Tabor House, Norwich Road, Malbarton, Norwich.
Meetings: The Observatory, Colney Lane, Colney, Norwich. Every Friday, 7.30 p.m.

Nottingham Astronomical Society
Secretary: C. Brennan, 40 Swindon Close, Giltbrook, Nottingham.

Oldham Astronomical Society
Secretary: P. J. Collins, 25 Park Crescent, Chadderton, Oldham.
Meetings: Werneth Park Study Centre, Frederick Street, Oldham. Fortnightly, Friday.

Open University Astronomical Society
Secretary: Jim Lee, c/o above, Milton Keynes.
Meetings: Open University, Walton Hall, Milton Keynes. As arranged.

Orpington Astronomical Society
Secretary: Miss Lucinda Jones, 263 Crescent Drive, Petts Wood, Orpington, Kent BR5 1AY.
Meetings: Orpington Parish Church Hall, Bark Hart Road. Thursdays monthly, 7.30 p.m. Sept.–July.

Plymouth Astronomical Society
Secretary: Sheila Evans, 40 Billington Close, Eggbuckland, Plymouth.
Meetings: Glynnis Kingdon Centre. 2nd Friday each month.

Portsmouth Astronomical Society
Secretary: G. B. Bryant, 81 Ringwood Road, Southsea.
Meetings: Monday. Fortnightly.

Preston & District Astronomical Society
Secretary: P. Sloane, 77 Ribby Road, Wrea Green, Kirkham, Preston, Lancs.
Meetings: Moor Park (Jeremiah Horrocks) Observatory, Preston. 2nd Wednesday. Last Friday each month. 7.30 p.m.

The Pulsar Group
Secretary: Barry Smith, 157 Reridge Road, Blackburn, Lancs.
Meetings: Amateur Astronomy Centre, Clough Bank, Bacup Road, Todmorden, Lancs. 1st Thursday each month.

Reading Astronomical Society
Secretary: Mrs Muriel Wrigley, 516 Wokingham Road, Earley, Reading.
Meetings: St Peter's Church Hall, Church Road, Earley. Monthly (3rd Sat.), 7 p.m.

Renfrew District Astronomical Society (formerly Paisley A.S.)
Secretary: Robert Law, 14d Marmion Court, Forkes, Paisley.

Richmond & Kew Astronomical Society
Secretary: Emil Pallos, 10 Burleigh Place, Cambalt Road, Putney, London SW15.
Meetings: Richmond Central Reference Library, Richmond, Surrey.

Salford Astronomical Society
Secretary: J. A. Handford, 45 Burnside Avenue, Salford 6, Lancs.
Meetings: The Observatory, Chaseley Road, Salford.

Salisbury Astronomical Society
Secretary: Mrs R. Collins, Mountains, 3 Fairview Road, Salisbury, Wilts.
Meetings: Salisbury City Library, Market Place, Salisbury.

Sandbach Astronomical Society
Secretary: Phil Benson, 8 Gawsworth Drive, Sandbach, Cheshire.
Meetings: Sandbach School, as arranged.

Scarborough & District Astronomical Society
Secretary: D. M. Mainprize, 76 Trafalgar Square, Scarborough, N. Yorks.
Meetings: Scarborough Public Library. Last Saturday each month, 7–9 p.m.

Scottish Astronomers Group
Secretary: G. Young c/o Mills Observatory, Balgay Park, Ancrum, Dundee.
Meetings: Bi-monthly, around the Country. Syllabus given on request.

Sheffield Astronomical Society
Secretary: Mrs Lilian M. Keen, 21 Seagrave Drive, Gleadless, Sheffield.
Meetings: City Museum, Weston Park, 3rd Friday each month. 7.30 p.m.

Sidmouth and District Astronomical Society
Secretary: M. Grant, Salters Meadow, Sidmouth, Devon.
Meetings: Norman Lockyer Observatory, Salcombe Hill. 1st Monday in each month.

Solent Amateur Astronomers
Secretary: R. Smith, 16 Lincoln Close, Woodley, Romsey, Hants.
Meetings: Room 2, Oaklands Community Centre, Fairisle Road, Lordshill, Southampton. 3rd Tuesday.

Southampton Astronomical Society
Secretary: C. R. Braines, 1a Drummond Road, Hythe, Southampton.
Meetings: Room 148, Murray Building, Southampton University, 2nd Thursday each month, 7.30 p.m.

South Astronomical Society
Secretary: G. T. Elston, 34 Plummer Road, Clapham Park, London SW4 8HH.

South Downs Astronomical Society
Secretary: J. Green, 46 Central Avenue, Bognor Regis, West Sussex.
Meetings: Assembly Rooms, Chichester. 1st Friday in each month.

South East Essex Astronomical Society
Secretary: C. Jones, 92 Long Riding, Basildon, Essex.
Meetings: Lecture Theatre, Central Library, Victoria Avenue, Southend-on-Sea. Generally 1st Thursday in month, Sept.–May.

South-East Kent Astronomical Society
Secretary: P. Andrew, 7 Farncombe Way, Whitfield, nr. Dover.
Meetings: Monthly.

South Lincolnshire Astronomical & Geophysical Society
Secretary: G. T. Walker, 7 Nook Lane, Empingham, Oakham, Leics.
Meetings: South Holland Centre, Spalding. 3rd Thursday each month, 7.30 p.m.

South London Astronomical Society
Chairman: P. Bruce, 2 Constance Road, West Croydon CR0 2RS.
Meetings: Surrey Halls, Birfield Road, Stockwell, London SW4. 2nd Tuesday each month, 8 p.m.

Southport Astronomical Society
Secretary: R. Rawlinson, 188 Haig Avenue, Southport, Merseyside.
Meetings: Monthly Sept.–May, plus observing sessions.

Southport, Ormskirk and District Astronomical Society
Secretary: J. T. Harrison, 92 Cottage Lane, Ormskirk, Lancs L39 3NJ.
Meetings: Saturday evenings, monthly as arranged.

South Shields Astronomical Society
Secretary: c/o South Tyneside College, St George's Avenue, South Shields.
Meetings: Marine and Technical College. Each Thursday, 7.30 p.m.

South Somerset Astronomical Society
Secretary: G. McNelly, 11 Laxton Close, Taunton, Somerset.
Meetings: Victoria Inn, Skittle Alley, East Reach, Taunton. Last Saturday each month, 7.30 p.m.

South West Cotswolds Astronomical Society
Secretary: C. R. Wiles, Old Castle House, The Triangle, Malmesbury, Wilts.
Meetings: 2nd Friday each month, 8 p.m. (Sept.–June).

South West Herts Astronomical Society
Secretary: Frank Phillips, 54 Highfield Way, Rickmansworth, Herts.
Meetings: Rickmansworth. Last Friday each month, Sept.–May.

Stafford and District Astronomical Society
Secretary: Mrs L. Hodkinson, Beecholme, Francis Green Lane, Penkridge, Staffs.
Meetings: Riverside Centre, Stafford. Every 3rd Thursday, Sept.–May, 7.30 p.m.

Stirling Astronomical Society
Secretary: R. H. Lynn, 25 Pullar Avenue, Bridge of Allan, Stirling.
Meetings: Smith Museum & Art Gallery, Dumbarton Road, Stirling. 2nd Friday each month, 7.30 p.m.

Stoke-on-Trent Astronomical Society
Secretary: M. Pace, Sundale, Dunnocksfold Road, Alsager, Stoke-on-Trent.
Meetings: Cartwright House, Broad Street, Hanley. Monthly.

Sussex Astronomical Society
Secretary: Mrs C. G. Sutton, 75 Vale Road, Portslade, Sussex.
Meetings: English Language Centre, Third Avenue, Hove. Every Wednesday, 7.30–9.30 p.m. Sept.–May.

Swansea Astronomical Society
Secretary: G. P. Lacey, 32 Glenbran Road, Birchgrove, Swansea.
Meetings: Dillwyn Llewellyn School, John Street, Cockett, Swansea. Second and fourth Thursday each month at 7.30 p.m.

Tavistock Astronomical Society
Secretary: D. S. Gibbs, Lanherne, Chollacott Lane, Whitchurch, Tavistock, Devon.
Meetings: Science Laboratory, Kelly College, Tavistock. 1st Wednesday in month. 7.30 p.m.

Thames Valley Astronomical Group
Secretary: K. J. Pallet, 82a Tennyson Street, South Lambeth, London SW8 3TH.
Meetings: Irregular.

Thanet Amateur Astronomical Society
Secretary: P. F. Jordan, 85 Crescent Road, Ramsgate.
Meetings: Hilderstone House, Broadstairs, Kent. Monthly.

Torbay Astronomical Society
Secretary: R. Jones, St Helens, Hermose Road, Teignmouth, Devon.
Meetings: Town Hall, Torquay. 3rd Thursday, Oct.–May.

Tullamore Astronomical Society
Secretary: S. Reynolds, Screggan, Tullamore, Co. Offaly, Ireland.
Meetings: Tullamore Vocational School, 1st Tuesday each month *and* VEC Office, High Street, last Friday of month. 8 p.m.

Usk Astronomical Society
Secretary: D. J. T. Thomas, 20 Maryport Street, Usk, Gwent.
Meetings: Usk Adult Education Centre, Maryport Street. Weekly, Thursdays (term dates).

Vectis Astronomical Society
Secretary: J. W. Smith, 27 Forest Road, Winford, Sandown, I.W.
Meetings: 4th Friday each month, except Dec. at Lord Louis Library Meeting Room, Newport, I.W.

Warwickshire Astronomical Society
Secretary: R. D. Wood, 20 Humber Road, Coventry, Warwickshire.
Meetings: 20 Humber Road, Coventry. Each Tuesday.

Webb Society
Secretary: S. J. Hynes, 8 Cormorant Close, Sydney, Crewe, Cheshire.
Meetings: As arranged.

Wellingborough District Astronomical Society
Secretary: S. M. Williams, 120 Brickhill Road, Wellingborough, Northants.
Meetings: On 2nd Wednesday. Gloucester Hall, Church Street, Wellingborough, 7.30 p.m.

Wessex Astronomical Society
Secretary: Leslie Fry, 14 Hanhum Road, Corfe Mullen, Dorset.
Meetings: Allendale Centre, Wimborne, Dorset. 1st Tuesday of each month.

West of London Astronomical Society
Secretary: A. H. Davis, 49 Beaulieu Drive, Pinner, Middlesex HA5 1NB.
Meetings: Monthly, alternately at Hillingdon and North Harrow. 2nd Monday of the month, except August.

West Midland Astronomical Association
Secretary: Miss S. Bundy, 93 Greenridge Road, Handsworth Wood, Birmingham.
Meetings: Dr Johnson House, Bull Street, Birmingham. As arranged.

West Yorkshire Astronomical Society
Secretary: K. Willoughby, 11 Hardisty Drive, Pontefract, Yorks.
Meetings: Rosse Observatory, Carleton Community Centre, Carleton Road, Pontefract, each Tuesday, 7.15 to 9 p.m.

Whittington Astronomical Society
Secretary: Peter Williamson, The Observatory, Top Street, Whittington, Shropshire.
Meetings: The Observatory every month.

Wolverhampton Astronomical Society
Secretary: M. Astley, Garwick, 8 Holme Mill, Fordhouses, Wolverhampton.
Meetings: Beckminster Methodist Church Hall, Birches Road, Wolverhampton. Alternate Mondays, Sept.–April.

Worcester Astronomical Society
Secretary: Arthur Wilkinson, 179 Henwick Road, St Johns, Worcester.
Meetings: Room 117, Worcester College of Higher Education, Henwick Grove, Worcester. 2nd Thursday each month.

Worthing Astronomical Society
Contact: G. Boots, 101 Ardingly Drive, Worthing, Sussex.
Meetings: Adult Education Centre, Union Place, Worthing, Sussex. 1st Wednesday each month (except August). 7.30 p.m.

Wycombe Astronomical Society
Secretary: P. A. Hodgins, 50 Copners Drive, Holmer Green, High Wycombe, Bucks.
Meetings: 3rd Wednesday each month, 7.45 p.m.

York Astronomical Society
Secretary: Simon Howard, 20 Manor Drive South, Acomb, York.
Meetings: Goddricke College, York University. 1st and 3rd Fridays.

Any society wishing to be included in this list of local societies or to update details are invited to write to the Editor (c/o Messrs Sidgwick & Jackson (Publishers), Ltd, Cavaye Place, London SW10 9PG), so that the relevant information may be included in the next edition of the *Yearbook*.